Modern and Contemporary Poetry and Poetics

Series Editor
Ann Vickery, Deakin University, Burwood, Australia

Founded by Rachel Blau DuPlessis, continued by David Herd, and now headed by Ann Vickery, *Modern and Contemporary Poetry and Poetics* promotes and pursues topics in the burgeoning field of 20th and 21st century poetics. Critical and scholarly work on poetry and poetics of interest to the series includes: social location in its relationships to subjectivity, to the construction of authorship, to oeuvres, and to careers; poetic reception and dissemination (groups, movements, formations, institutions); the intersection of poetry and theory; questions about language, poetic authority, and the goals of writing; claims in poetics, impacts of social life, and the dynamics of the poetic career as these are staged and debated by poets and inside poems. Since its inception, the series has been distinguished by its tilt toward experimental work – intellectually, politically, aesthetically. It has consistently published work on Anglophone poetry in the broadest sense and has featured critical work studying literatures of the UK, of the US, of Canada, and Australia, as well as eclectic mixes of work from other social and poetic communities. As poetry and poetics form a crucial response to contemporary social and political conditions, under David Herd's editorship the series will continue to broaden understanding of the field and its significance.

Paul Keers

T.S. Eliot

A Lifetime of Old Age

palgrave
macmillan

Paul Keers
London, UK

ISSN 2634-6052 ISSN 2634-6060 (electronic)
Modern and Contemporary Poetry and Poetics
ISBN 978-3-032-25049-0 ISBN 978-3-032-25050-6 (eBook)
https://doi.org/10.1007/978-3-032-25050-6

Cover credit: © Harvey Loake

This Palgrave Macmillan imprint is published by the registered company Springer Nature Switzerland AG
The registered company address is: Gewerbestrasse 11, 6330 Cham, Switzerland

Acknowledgements My thanks are due to all those who encouraged me to embark upon this study. Initial reassurance that it was a subject worthy of pursuit came from Professor Seamus Perry of Balliol College, Oxford, and in particular from my former tutor Professor Stephen Gill of Lincoln College, Oxford. Their encouragement at the outset was invaluable.

It was thanks to Bill Trent that I pursued this as a book, rather than as postgraduate research, while Hugh Black-Hawkins and the Rev Dr. John Caperon provided valuable guidance on its direction. Alongside John, I must also thank Kathy Radley and the other members of the Committee of the T.S. Eliot Society (UK), who granted me leeway when my work on the book distracted me from my responsibilities as Chair.

Thanks are due to Clare Reihill at the Eliot Estate for her support at the outset, and to Becky Taylor, Permissions Manager at Faber & Faber, Eliot's publishers, for her guidance on their fair dealing policy which governs the study's use of quotations from his works. The equivalent policies of HarperCollins US and Yale University Press govern the use of quotations to which they control US rights.

I am grateful to the staff of the Senate House Library, London, and the Bodleian Library, Oxford for their cheerful accommodation of my requirements, and to Eileen Srebernik, Senior Editor at Palgrave, together with my peer reviewer, for refining my initial submission.

Finally, thanks to my beloved wife Carrie who, even when building work meant that I had to appropriate her own study, encouraged, supported and tolerated me, not only throughout my immersion in this book, but through life in general. Thanks to her, I can say with Eliot that in later life I too "have been brushed by the wing of happiness".

Competing Interests The author has no competing interests to declare that are relevant to the content of this manuscript.

Contents

1 Introduction 1
2 "I Was Born Old": 1905–1915 13
3 "I Grow Old, I Grow Old": 1916–1921 29
4 "Maturity of Mind" 45
5 "I Tiresias, Old Man": 1922–1926 63
6 "The Agèd Eagle": 1927–1935 83
7 "The Wisdom of Old Men": 1936–1946 103
8 "An Elder Statesman": 1947–1965 127

Index 147

CHAPTER 1

Introduction

Abstract This study seeks to establish the significance of old age as a theme running throughout Eliot's work. In addition to works whose titular subjects are old men, more than thirty of his poems contain references to old age and ageing. His poetry, plays and critical prose all employ old age as a device, while for most of his life, Eliot himself projected an age and maturity greater than his actual years. The study traces an arc from his youthful depictions of decrepitude in old age; through a period, following his religious conversion, of elderly characters turning unfulfilled towards death; to a later expression of his personal and emotional experience of senescence. It also examines the incorporation of his attitudes towards age into his critical approach, his use of 'maturity' and 'immaturity' as critical criteria, and the intersections between age, tradition, tribute and legacy in his works.

Keywords Old age · Modernist portrayals · Projection · Age studies · Summary of content

T. S. Eliot was much possessed by age, and often saw an agèd skull beneath the skin. His *Collected Poems* contains three works—'Gerontion', 'A Song for Simeon' and 'Lines for an Old Man'—whose titular subjects are old men. Closer inspection reveals his poetry to be populated with

P. Keers, *T.S. Eliot*, Modern and Contemporary Poetry and Poetics,
https://doi.org/10.1007/978-3-032-25050-6_1

elderly characters (such as Tiresias, Phlebas, the Magus, Saint Narcissus and Marie), with over thirty poems containing references to ageing or old age.

Eliot adopts the soubriquets 'agèd eagle' and 'Old Possum' for himself, and accepts from colleagues the nickname 'The Elephant', an animal associated with old age, memory and wisdom. He projects an age beyond his actual years; and he uses youth and maturity as critical criteria. He then emerges from behind the personae to consider in the first person his own senescence in *Four Quartets*. Finally, he employs old age as a theme in his dramas, first in *The Family Reunion* and then *The Elder Statesman*, with the latter title reflecting his own status in later life.

From his earliest juvenilia to his last Ariel poem, Eliot uses the reminiscences of an old man as a poetic device. Old age is present in the epigraphs to his poems, the judgements in his critical prose, and the assessments in his memorial tributes. We can trace a lifetime of engagement with old age in his poetry, prose, plays and public life, and an abiding preoccupation with old age in his letters. "I resigned myself to celibate old age"[1] he wrote in his early forties; yet finally he contradicts his lifelong apprehension with the declarations of sensuality aged sixty-nine in 'A Dedication to my Wife'.

This study seeks to establish the significance of old age as a theme in Eliot's work. Throughout his lifetime and across his literary output, Eliot returned to that single concept—old age. Narrators, male and female characters, even animals, are presented either implicitly or explicitly as old; indeed several characters are described simply as an 'old man'. Elderly characters like Hieronymo are taken from historical sources, while Eliot transforms into old men adopted figures whose ages are not defined in their sources. Observations, even while Eliot is objectively youthful, are made through the perspective of the elderly. The preponderance of references to, by or through agèd personae is marked.

In the period Eliot would describe as *l'entre deux guerres*, several of his Modernist contemporaries addressed the subject of old age, displaying an anxiety about senescence with often unsympathetic portrayals. The final volume of Marcel Proust's *À la recherche du temps perdu* (1927) concludes with a party at which old age is pictured as "repulsive".[2] The characters from his novel are unrecognisably transformed in old age, with "disfigurement which offended my eyes".[3] Men who, "beneath their hood of white hair, had already the rigidity, the sealed eyelids of those who are about to die",[4] resemble "one of those fine antique heads, eaten away and hopelessly damaged"[5]; the faces of women, who "strained every nerve in a ceaseless struggle against old age",[6] become "a terrible ossified mask".[7] Limping, bowed or corpulent, many of the narrator's acquaintances can still be recognised in their old age, "but only as rather bad portraits of themselves".[8] While the party inspires the narrator's perceptions of Time and memories, he mourns in this encounter with old age "the extinction of youth, the destruction of a person full of energy and high spirits".[9] Although "proud to have introduced to English readers the work of Marcel Proust"[10] (in the *Criterion* of July 1924), Eliot felt that Proust was "far too much a sensationalist"[11]; nevertheless, his images of old age resonate with Eliot's own.

Similarly, in the narrative of F. Scott Fitzgerald's *The Beautiful and Damned* (1922), time plays "its ultimate chuckling tricks",[12] and ageing and anxiety about old age is a central theme. The novel's main character, Anthony Patch, awaits the inheritance he anticipates on the death of his grandfather, whom he regards as a "feeble, unintelligent old man"[13] and "a back number now"[14] despite his wealth and power. Anthony himself seeks "some path of hope to lead him toward what he thought was an imminent and ominous old age".[15] Anthony's wife, Gloria, shares this despair at growing old, which she describes as "like begging for mercy from a storm".[16]

For Fitzgerald's characters, ageing begins in what most would regard as youth. Anthony's friend Maury declares:

> "We're growing old, Anthony. I'm twenty-seven, by God! Three years to thirty, and then I'm what an undergraduate calls a middle-aged man."
> "You are old, Maury," [Anthony] agreed at length. "The first signs of a very dissolute and wabbly senescence".[17]

This perspective is retained in *The Great Gatsby* (1925), a book which Eliot said in a letter to Fitzgerald "has interested and excited me more than any new novel I have seen, either English or American, for a number of years".[18] There, narrator Nick Carraway laments his thirtieth birthday in distinctly Prufrockian manner:

> "Thirty — the promise of a decade of loneliness, a thinning list of single men to know, a thinning briefcase of enthusiasm, thinning hair."[19]

Taking a more sympathetic view of old age, and perhaps also conscious of the post-War literary focus upon young men, Virginia Woolf declares in *Mr Bennett and Mrs Brown* (1924) that the central character of a modern novel by a Georgian writer might well be an elderly woman. "I believe that all novels begin with an old lady in the corner opposite"[20] she writes, and proceeds to elaborate upon the imagined character and situation of 'Mrs Brown', as opposed to a didactic emphasis upon what Woolf describes as "the fabric of things".[21] This reflects her "curiosity about human beings" to which Eliot refers in *Horizon* magazine[22] after her death (Chapter 8).

And in 1931 the young Samuel Beckett, then at Trinity College, Dublin, conceived and himself performed in a parody of Corneille's *Le Cid*, which he called *Le Kid*. "Beckett, as Le Kid's father, Don Diegue, portrayed a very old man with a long white beard holding a small alarm clock in his hands".[23] As with Eliot, an old man is appearing in the writer's youthful work; the description also matches that of the "Old Gentleman" Eliot adds in to *Sweeney Agonistes* in March 1933, who "resembles closely Father Christmas" and carries "an alarum [*sic*] clock" (See Chapter 6). Of all the writers of this period, elderly characters would most significantly dominate Beckett's subsequent work, in which, as Elizabeth Barry notes, "Old age—or at least the tropes associated with it—are everywhere, and Beckett seemed to appropriate the condition as spiritually his own".[24]

While Eliot reflects many of his contemporaries' anxieties, it is with the declarations of old age in the late poems of W.B. Yeats with which he has the most significant relationship. After years of criticising Yeats, Eliot declared in 1933 that Yeats's late work contains "some of the most beautiful poetry in the language, some of the clearest, simplest, most direct".[25] He later observes Yeats "experiencing new emotions appropriate to one's age, and with the same intensity as the emotions of youth".[26] Of Yeats's candid expressions of old age, Eliot asks "To what honest man, old enough, can these sentiments be entirely alien?",[27] and Yeats becomes one of the figures comprising the "compound ghost" who lectures Eliot upon old age in 'Little Gidding'. As we shall see in a lengthier consideration in Chapter 7, this acknowledgement only came when Eliot himself had reached the age at which Yeats had composed his later poetry.

Eliot 'felt old' from an early age; indeed he once wrote that he was "born old".[28] His fellow poet C. Day Lewis wrote that "It is generally accepted that Eliot as a poet was born middle-aged, and has not been getting any younger since".[29] Eliot wrote the famous line "I grow old, I grow old" in 'The Love Song of J. Alfred Prufrock' at the age of twenty-one; he complained at twenty-five that "I am so old that it makes me despair"[30]; and he received criticism from the critic Edmund Wilson for presenting himself as an "agèd eagle" at the age of forty-one.[31] As Kathleen Woodward points out, "Although Eliot certainly possesses an acute sense of history, his sense of time in terms of his own life is severely impoverished".[32]

John Soldo proposes that "This disguise of being older than one is is part of the overall mask Eliot adopted",[33] and that this denial of his past accords with Eliot's pursuit of impersonality in his poetry and maturity in his prose. We are constantly drawn away from his own identity and age into another persona, often that of an older man. This is certainly the case in much of Eliot's poetry; Martin Scofield has similarly noted Eliot's "creation of a narrator-mask", "avoiding direct utterance and speaking through masks of personae", which "avoids the direct autobiographical utterance", and hence "is in many ways the corollary of the idea of impersonality".[34]

But Soldo goes beyond the deployment of a mask as a poetic device, and suggests that "the mask is a reflection of the struggle of the personality to create a social identity".[35] This does not acknowledge its persistence long after he had achieved literary status, established his social identity and indeed reached an age at which he was objectively an old man. Similarly, the idea that this is a mask or disguise implies its conscious adoption, whereas its presence throughout Eliot's corpus, coupled with references to old age in his prose, his personal conversations and letters, and his interviews, combine to suggest a lifelong compelling and besetting concern.

Related to his adoption of a greater age is the emphasis which he places upon 'maturity' as a criterion in his critical writings. Throughout his career, he identifies and praises writers for exhibiting maturity, and correspondingly criticises those guilty of what he perceives to be immaturity, youthfulness and adolescence. Although these terms are commonly associated with the passage of lifetimes, Eliot asserted maturity for himself, including 'old age', while in his twenties. "The polemical rejection of any developmental understanding of self", writes David Rosen, "required both a forceful insistence on the poet's own maturity and a concealment of his past".[36] By identifying the characteristics by which Eliot observes maturity in others, we see in Chapter 4 how Eliot applied them to himself and displayed them in his work, thereby exhibiting as a young man the 'maturity' he venerated.

This study appears at a time when the discipline of Age Studies, and the role of literature within it, has become an increasingly significant academic field. Age Studies focusses primarily upon the relevance of representations of old age to society. When Marion Gymnich explores 'The Value of Literature in an Aging Society', she "starts from the assumption that literature may play a vital role in cultural negotiations of old age and aging by creating a deeper understanding of and empathy for the situation of elderly people".[37]

Literary studies may generate material for such "cultural negotiations", but that is not their primary objective. As Kathleen Woodward makes clear in *At Last, The Real Distinguished Thing: The Late Poems of Eliot, Pound, Stevens and Williams*:

> I did not begin my work with questions about the "images" of poets in their old age. If I had, I might have written a book dominated by images of aging that would have sought, consciously, to champion the rights of the elderly as a dispossessed group, much as books have been written on behalf of women and their literature, blacks and their literature.[38]

Instead, Woodward set out to ask the literary questions: "Did aging and old age bring poetic fulfilment? Do these poets offer us insight into the experience of aging and its satisfactions?"[39] These are questions of relevance to a study of Eliot.

Nevertheless, the cultural representations of old age surfacing in Age Studies offer an interesting perspective from which to view Eliot's own deployment of old age. In 'Ageing in Poetry: A Windfall', her chapter in *The Bloomsbury Handbook to Ageing in Contemporary Literature and Film*,[40] Tess Maginnis writes that "There is still a tendency to polarize perceptions of ageing in cultural discourses as broadly following a decline narrative or, conversely, a narrative of successful ageing". She cites a corresponding formulation by Aagje Swinnen of stereotypical conceptualisations of age as "opposing discourses of physical and mental decline versus wisdom and transcendence".[41] Or as Woodward summarises,

> The problem, or weakness, of research on aging and the elderly is that it tends to fall too easily into one of two overly simplified categories: either aging and old age are romanticized or they are presented as tragic

> twentieth-century injustices that can be alleviated, perhaps even rectified, through medical research and political and social reform.[42]

Eliot's conceptualisations of age do not align with either simple narrative of ageing. Old age is used by Eliot as a fluid concept, varying from the negative, critical and sometimes repulsive, to the positive, respectful and even affectionate. Imagery of geriatric physical deterioration and decline appear in his most youthful work; in his mature works he subverts the stereotypical conceptualisations of successful ageing by denying the wisdom of old men, and dismissing the value of experience. He also ultimately declares the sensual nature and physical vigour of his senescent marital relations. His portrayals of old age therefore present us with both decline and success, physical and mental, temporal and spiritual, often together in single characters, in a complex presentation of senescence.

There is also the issue that Eliot did not write his poetry as an old man himself. This is not necessarily a creative handicap; as Richard A. Posner observes in *Aging and Old Age*, a work which examines social, biological, economic and normative issues of ageing, "'Gerontion', published when [Eliot] was in his early thirties, shows that you don't have to be old to write perceptively about old age".[43] But Age Studies primarily explores the expression in literature of the *experience* of old age, and as Woodward acknowledges, when asking the literary questions quoted above, while her other poets produced their 'last' works in their sixties and seventies, Eliot's *Four Quartets* was published in his fifties. "But its voice and theme are definitely that of an older man".[44] She observes that:

> It is crucial for us [...] to project ourselves imaginatively into our own future, our old age, for as individuals we tend to repress old age psychologically just as industrial culture has devalued the elderly as a nonproductive group and pushed them out of sight. But the surfacing of old age, in Eliot's sense, in middle age is a chilling prospect [...] Eliot imposes a dark vision of old age on his middle years. He is, we would say, old before his time, and here we mean old in the pejorative sense.[45]

Tracing the presence and purpose of old age across Eliot's *oeuvre*, this study has an approximately chronological structure. Chapter 2 examines Eliot's poems of childhood and youth, which contain for a young writer a significant number of images of old age, several of which will recur in his later work, and the adoption of poetic devices such as projection into an older persona. His youthful poetry displays a tension between youth and age, reflecting the conflicting influences of the principle of lifelong work and duty espoused by his grandfather, Rev. William Greenleaf Eliot, and the *carpe diem* philosophy which Eliot encountered as a teenager in *The Rubáiyát of Omar Khayyám*. Eschewing associations with status and wisdom, he portrays old age as a deterioration both physical and mental, a senescence he appears to anticipate with dread.

This derogatory portrayal of old age persists through Eliot's first published poetry. The age of J. Alfred Prufrock himself is indeterminate, but shown to be older than his author at the time, while lines omitted from the poem in 'Prufrock's Pervigilium' combine aspects of Eliot's previous images of old men into a single character. The poems which follow in Chapter 3 present senescence as physically repulsive and sexually aberrant, until the period culminates in 'Gerontion', a psychologically sophisticated (and clinically accurate) portrayal of old age with an old man as narrator embodying loss, failure and decline.

Throughout his career, Eliot's criticism ran in parallel to his poetic and, ultimately, dramatic writing. Removed from the chronological arc of his creative writing, Chapter 4 considers Eliot's critical prose across his lifetime, and the ways in which it incorporates his attitudes towards age, particularly in his efforts to position himself away from adolescence and youth. He creates an associated critical framework around concepts of 'maturity' and 'immaturity'; 'maturity' is reserved for the most successful and acclaimed works, and, by implication, for those by Eliot himself, who claims and insists from an early age upon his own maturity. 'Mature artists steal', he writes, simultaneously justifying his emerging technique of allusion, and claiming maturity for himself.

Old age pervades *The Waste Land*, the work which dominates Chapter 5. Its central figure, Tiresias, is presented by Eliot as an old man and, according to Eliot, unites the poem's other characters. These, from the Sibyl of the epigraph to Hieronymo in the final passage, exhibit characteristics of old age including failing eyesight, insomnia, voyeurism, deafness, dementia and despair; they often remember but rarely act. Each illuminates the poem through the perspective of their age, and tension between these elderly characters and the activities of the young is a significant aspect of the poem. This period concludes with 'The Hollow Men', a poem whose mood of despair is intensified by portraying its characters as in their final years.

Eliot's religious conversion in 1927 leads to the shift in his portrayal of old age seen in Chapter 6. In his first two Ariel poems, he employs the elderly biblical figures of Simeon and the Magi in order to address the new perspectives which his conversion presented, and to contrast those both with the elderly characters he had previously depicted, and with the philosophy of his grandfather. In 'Ash-Wednesday' he leaves behind the negative image of old age which he has previously deployed, and instead presents himself in his forties as an "agèd eagle", a position of seniority, from which he can castigate another ageing literary figure's behaviour in 'Lines for an Old Man'.

The rarely-considered significance of a warm and avuncular Old Possum, whose cats exhibit positive aspects of old age and retirement, is considered in Chapter 7, alongside the contemporaneous drama *The Family Reunion*, whose elderly central characters embody issues of the futility of temporal action, and the value of late spiritual insight. Eliot now acknowledges the honesty and universal truths in the observations upon old age in the late poems of W.B. Yeats. And *Four Quartets*, published together while Eliot was only in his mid-fifties, subsequently contains a powerful first-person assessment of the old age Eliot feels that he is now experiencing for himself, detailing failings and 'wrongs' in old men and in his own perceived dotage.

It was only as his first wife, Vivienne, was lowered into her grave that he claimed he became his chronological age, then fifty-eight.[46] In his final decades, considered in Chapter 8, Eliot wrote many eulogies and tributes for passing members of his generation; these are often more personal than his other writings, and reflect both his assessment of the value in a lifetime's work, and a desire in his own late years to establish his legacy. His last Ariel poem returns to his poetic device of an old man's reminiscence; and his final play, *The Elder Statesman*, presents a titular character reflecting not only Eliot's public persona, but his personal concerns about his past and his late realisation of the power of human love. In the play's dedication to his second wife, and its subsequent version as a poem, Eliot celebrates his unexpected discovery of love and sensual passion aged sixty-eight, repudiating his lifelong vision of senescence, and concluding in life the old age which he had presented with such negativity in unanticipated joy and contentment.

In 1951, an article by Gordon Symes on 'Eliot and Old Age' noted the presence of the theme in some of his major works. Later scholarly work focussing on individual poems and periods, cited in the relevant chapters, confirmed its presence elsewhere. These studies have been invaluable in establishing its significance across his canon, while the recent authorised biographies, and the editing and ongoing publication of Eliot's letters, including the release of his letters to Emily Hale, have revealed his life-long personal concern about ageing. The broadly chronological structure of this study should not be taken to imply that there is a 'development' in Eliot's deployment of old age (and his critical work is considered outside this chronology). However, an arc can certainly be delineated from the youthful influence of a *carpe diem* philosophy, with accompanying views of decrepitude in old age; through a period, following his religious conversion, of elderly characters turning towards death; to a later examination of the personal and emotional experience of old age. It is both ironic and gratifying that Eliot's final expressions of a temporally satisfying conclusion to life contradict many of the negative anticipations of old age contained within his lifetime's work.

Notes

1. Letter to Emily Hale, 3rd November 1930.
2. Proust, Marcel. 1981. *Remembrance of Things Past*. Translated by C. K. Scott Moncrieff and Terence Kilmartin, Chatto & Windus 3, 988.
3. Ibid., 975.
4. Ibid., 981.
5. Ibid., 1070.
6. Ibid., 989.
7. Ibid., 1049.
8. Ibid., 978.
9. Ibid., 983.
10. Eliot, T. S. 2022. *The Complete Prose of T. S. Eliot: The Critical Edition Vols 1–8*. Ed. Schuchard, Ronald, Brooker, Jewel Spears, Cuda, Anthony, Dickey, Frances, Formichelli, Jennifer, Harding, Jason, Javadi, Iman, Stayer, Jayme and Chinitz, David. The Johns Hopkins University Press and Faber & Faber Ltd. http://muse.jhu.edu/resource/1 8, 570.
11. Eliot, T.S. 2009–2025. *The Letters of T.S. Eliot Vols 1–10* Ed. Valerie Eliot, Hugh Haughton and John Haffenden. Faber & Faber, 2, 356.

12. Fitzgerald, F. Scott. 1922. *The Beautiful and Damned*. Charles Scribner's Sons, 138.
13. Ibid., 14.
14. Ibid., 223.
15. Ibid., 55.
16. Ibid., 360.
17. Ibid., 50.
18. Eliot, *Letters* 2, 813.
19. Fitzgerald, F. Scott. 1925. *The Great Gatsby*. Charles Scribner's Sons, Scribner Classic/Collier Edition, 1986, 136.
20. Woolf, Virginia. 1924. *Mr Bennett and Mrs Brown*. Hogarth Press, 9.
21. Ibid., 18.
22. Eliot, *Prose* 6, 169.
23. Bair, Deirdre. 1978. *Samuel Beckett: A Biography*, Jonathan Cape, 127.
24. Barry, Elizabeth. 2016. 'Samuel Beckett and the Contingency of Old Age', *Samuel Beckett Today / Aujourd'hui*, Vol. 28, No. 2, Special Issue: Clinique et poétique du vieillir dans le théâtre de Beckett / Clinics and Poetics: Beckett's Theatre and Aging, 206.
25. Eliot, *Prose* 4, 679.
26. Eliot, *Prose* 6, 78.
27. Eliot, *Prose* 6, 84.
28. T.S. Eliot, undated leaf amomg TSE's letters to McKnight Kauffer, cited in Eliot, T.S. 2015. *The Poems of T.S. Eliot Vols I and II*. Ed. Christopher Ricks and Jim McCue, Faber and Faber, 1, 397.
29. Lewis, C. Day. 1939. *A Hope for Poetry*. Basil Blackwell, 18.
30. Patmore, Brigit. 1968. *My Friends When Young: The Memoirs of Brigit Patmore, edited and with an introduction by Derek Patmore*. William Heinemann, 89.
31. Wilson, Edmund, 1931, *Axel's Castle*, Charles Scribner's Sons, 130.
32. Woodward, Kathleen M. 1980. *At Last, the Real Distinguished Thing: the late poems of Eliot, Pound, Stevens, and Williams,* Ohio University Press, 58.
33. Soldo, John J. 1983. *The Tempering of T. S. Eliot*, UMI Research Press, 94.
34. Scofield, Martin. 1988. *T.S. Eliot: the poems*. Cambridge University Press, 76.

35. Soldo, 94.
36. Rosen, David. 2003, 'T. S. Eliot and the Lost Youth of Modern Poetry', *Modern Language Quarterly* 64:4, 473–494 https://doi.org/10.1215/00267929-64-4-473
37. Gymnich, Marion. 2020.'The Value of Literature in an Aging Society' REAL (Berlin, West), 2020-12, Vol.36 (1), 201.
38. Woodward, xi.
39. Ibid., 3.
40. Maginnis, Tess. 2023. 'Ageing in Poetry: A Windfall', *The Bloomsbury Handbook to Ageing in Contemporary Literature and Film*, ed. Sarah Falcus, et al., Bloomsbury Publishing, 76.
41. Swinnen, A. 2018. ' "Writing to Make Ageing New: Dutch Poets' Understandings of Late-Life Creativity', *Ageing & Society* 38, 543.
42. Woodward, xi.
43. Posner, Richard A. 1995. *Aging and Old Age*. University of Chicago Press, 100.
44. Woodward, 4.
45. Woodward, 56.
46. Eliot, T.S. Letter to Emily Hale, 14th February 1947.

CHAPTER 2

"I Was Born Old": 1905–1915

Abstract Eliot's youthful poems display negative characteristics of old age which persist and are subsequently deployed throughout his *oeuvre*. There is a consistent portrayal of old men as deteriorating, both physically and mentally; and after his initial graduation poems, there is an eschewal of the common notions that status and wisdom accrue with age. His early writings display evidence of the influences of his grandfather, Reverend William Greenleaf Eliot, who established his family's values of lifelong service, but also of *carpe diem* philosophies reflected in his readings of Omar Khayyám and Robert Louis Stevenson. Poetic devices to which Eliot will repeatedly return, such as projection into the perspectives and reminiscences of old men, and a tension between youth and age, are established, as are characteristics of senescence which will recur throughout his later work.

Keywords Juvenilia · Family · Grandfather · Stevenson · Khayyam · Death of St Narcissus · Pervigilium · Prufrock

It is not just the quantum of references and allusions to ageing in the poetry of a young man which is notable in T.S. Eliot's juvenilia. There is the emergence of negative characteristics of old age which will persist and be deployed throughout his *oeuvre*. There is a consistent portrayal

P. Keers, *T.S. Eliot*, Modern and Contemporary Poetry and Poetics, https://doi.org/10.1007/978-3-032-25050-6_2

of old men as deteriorating, both physically and mentally; and after his initial graduation poems, there is an eschewal of the common notions that status and wisdom accrue with age. Eliot began writing poetry as a schoolboy and his first poem to be printed was in the *Smith Academy Record* in 1905, written when he was sixteen. "Which is worse: a child writing as an adult or an adult writing as a child?" he was to wonder much later. "But as I was born old, I wouldn't know".[1] Eliot's juvenilia reflects two particularly significant and opposing influences—the philosophy of his grandfather, Reverend William Greenleaf Eliot, which was fundamental to his family upbringing; and his discovery of *The Rubáiyát of Omar Khayyám*, which he would later say caused his mind to burst into life at the age of fourteen.[2]

Although Eliot never personally knew William Greenleaf Eliot, who died a year before he was born, he later acknowledged that his grandfather provided the moral structure which governed his family upbringing, with a set of judgments upon duty, self-sacrifice and indulgence which dictated behaviour.[3] A Unitarian pastor and preacher, William Greenleaf Eliot established the first Unitarian church in St. Louis in 1834; he founded Washington University in 1853, and Smith Academy in 1854, the boys' school which Eliot would himself attend. A biography of William Greenleaf Eliot was written by Eliot's mother Charlotte and published in 1904 which carries the dedication, "Written for my children 'lest they forget'".[4] She recorded that he was described in a memorial sermon as "one who has led a steady life of self-denying usefulness".[5]

William Greenleaf Eliot's collection of lectures and sermons entitled *Lectures to Young Men*[6] included his last written sermon delivered in St Louis in 1885:

> in the days of youthful enjoyment and hope, let your service of God begin. The whole of your mortal life is not too great an offering to God. You cannot begin the work of self-consecrating devotion to duty too soon [...] Let no wild oats be sown, for be well assured a day of reaping will come. Your sin will, sooner or later, find you out [...]
> And for us who stand at the limit of the appointed years, or already beyond it, let us remember that while there is life there is work to do [...] The closing years of a faithful life-service are oftentimes the best. By a true spirit of self-sacrifice now, we may add force and strength to all that has gone before. The extended years [...] may possess more than youthful vigor, as the service of time is passing into the service of eternity.[7]

In fundamental opposition to the above passage is the *carpe diem* philosophy espoused in *The Rubáiyát of Omar Khayyám*, the collection of Persian poetry translated into English in the nineteenth century by Edward FitzGerald; Eliot later stated that his fourteen-year-old mind "burst into life on collision with Omar".[8] Immensely popular at the time, as Vinnie-Marie Ambrosio notes "*The Rubáiyát* had become, by the turn of the century, an object of admiration or attack on many levels, notably for its theology, for the politics that rose out of the theology, and for the call to free life that rose out of the politics".[9] This presented to the young Eliot, in an easily assimilable poetic form with powerful imagery, a Romantic alternative to the philosophies of his Puritan upbringing. Indeed, as he later admitted in a prize day speech to school pupils, the fact that its vision of youth, life and old age would have been disapproved of by his elders contributed to its appeal.[10]

The Rubáiyát suggests that "The Bird of Time has but a little way / To fly – and Lo! the Bird is on the Wing".[11] It encourages immediate satisfaction over long-term labours which will only conclude in physical death: "Ah, make the most of what we yet may spend / Before we too into the Dust descend".[12] William Greenleaf Eliot's "spirit of self-sacrifice" is countered with self-indulgence; where his sermon demands that "no wild oats be sown" with the threat of "a day of reaping", Khayyám rejects "the yoke / Of unpermitted Pleasure, under pain / Of Everlasting Penalties, if broke!"[13] *The Rubáiyát* not only rejects temperance and denies the ethic of hard work, but extols sensual pleasure, the waning of which in old age will diminish the purpose of life. These two philosophies of life and old age, the one instilled and the other discovered, will be grappled with throughout Eliot's subsequent canon.

It is in his school graduation poem 'To the Class of 1905'[14] that Eliot first employs the technique of projection into the perspective of an old man. He pictures his fellow students embarking into their futures with hope, eagerness and courage, travelling widely and performing great deeds. It is from this position of implied worldly success that they will return, in "distant years", old and with grey hair, to find a school changed in appearance, but not in values. This is an essentially positive vision of senescence, its achievements unquestioned, and the only physical deterioration in old age the greying of hair. The poem reflects the two philosophies which are influencing the young poet. It expresses both a desire to be one of those who try "To labor for the good until they die",[15] a clear reflection of William Greenleaf Eliot's mandate; but also an acknowledgement that the pupils' lives are "a momentary gleam / A bubble on the surface of the stream / A drop of dew upon the morning grass",[16] an expression of transience through natural imagery typical of lines in *The Rubáiyát* such as "I came like Water, and like Wind I go",[17] and "Like Snow upon the Desert's dusty Face, / Lighting a little hour or two – is gone".[18]

Five years later, graduating from Harvard, Eliot's 'Ode'(1910) employs the same trope of projection into the perspective of an older man. The years ahead, however, are now faced with hesitancy and fear; old age needs to be attained in order to appreciate the qualities which Harvard has provided, for "Only the years that efface and destroy / Give us also the vision to see".[19] The erosion which age inflicts is now being acknowledged, but is counterbalanced by an enhanced perception, a pairing of characteristics which Eliot will later grant to the wrinkled old man Tiresias. Nevertheless, enhanced perception is not a characteristic of old age which Eliot will often depict. Instead blindness (both literal and metaphorical) will be a more frequent trait of his elderly personae.

In 'Opera' (1909) Eliot positions himself as chronologically within, but perspectivally outside, the process of ageing:

> And I feel like the ghost of youth
> At the undertakers' ball[20]

The lines posit an existence situated between past youth and future death. The narrator, a representative of youth yet disconnected from it, foresees death in terms of a celebratory occasion. Reversing the common conceit of death as a prospect haunting youth, here youth is haunting death. Robert Crawford describes its tone as "in part young and in part beyond all youth".[21] It is a position which Eliot will increasingly adopt, wishing to demonstrate that he has left the youth and immaturity of his beginnings behind him.

It is not known when Eliot first read the essay 'Crabbed Age and Youth' by Robert Louis Stevenson, but he praised Stevenson in a 1909 undergraduate essay on Kipling[22] and would include Stevenson as an essayist in the syllabus for his course in Modern English Literature for the University of London in 1917. Citing the volume *Virginibus Puerisque*, in which the essay 'Crabbed Age and Youth' appeared,[23] its influence on Eliot would later be evidenced in images of age in both 'Rhapsody on a Windy Night' and 'The Love Song of J. Alfred Prufrock'.

'Crabbed Age and Youth' is a witty and impassioned argument for youthful *carpe diem*, incorporating depictions of mental and physical failings in old age. "It will not do", writes Stevenson, "to delay until we are clogged with prudence and limping with rheumatism".[24]

Stevenson echoes the youthful atheistic self-indulgence of the *Rubáiyát*:

> People may lay down their lives with cheerfulness in the sure expectation of a blessed immortality; but that is a different affair from giving up youth with all its admirable pleasures, in the hope of a better quality of gruel in a more than problematical, nay, more than improbable, old age. We should not compliment a hungry man, who should refuse a whole dinner and reserve all his appetite for the dessert, before he knew whether there was to be any dessert or not.[25]

Writing that "Old people have faults of their own; they tend to become cowardly, niggardly, and suspicious", Stephenson opines that old age leads to these and other faults, "Whether from the growth of experience or the decline of animal heat".[26] Eliot will challenge the supposed wisdom accrued from experience at several instances in *Four Quartets*; and sensuality in old age cools in 'Gerontion' and is expiring in 'Little Gidding'.

Whereas *The Rubáiyát*'s engagement with *carpe diem* is fuelled by the pleasures of youth, Stevenson's is driven by the prospect of an unpleasant old age. The two are in accord—as the one fades, the other approaches—and neither finds reconciliation in the prospect of an afterlife. Eliot drew poetic inspiration from *The Rubáiyát*; he later recalled that he began "to write a number of very gloomy and atheistical and despairing quatrains in the same style",[27] and 'A Lyric' (1905) extols a philosophy in which "our days of love be few".[28] Even at this early stage of Eliot's poetic output, his interpretation of *carpe diem* incorporates a questioning of the value of old age: "So why, Love, should we ever pray / To live a century?"[29] Rather than celebrating youth, a negative, derogatory vision of old age, closer to Stevenson's, underpins Eliot's subsequent imagery. For example, 'In the Department Store' (1915) summarises the brevity of life, a key theme of both Khayyám and Stevenson, with the bleak "Man's life is powerless and brief and dark".[30]

In the juvenilia eventually published after Eliot's death in *Poems Written in Early Youth*,[31] as well as in other unpublished poems of the period, Eliot presents a series of images of old men which are almost exclusively negative in their portrayal. In terms of the discourse of ageing narratives, Eliot can be seen to view old age not in terms of growth, but in terms of decline.

'First Debate between the Body and Soul' (1910) exists only in manuscript, but Eliot will draw again later upon its powerful image of:

> A blind old man who coughs and spits sputters
> Stumbling among the alleys and the gutters[32]

Rich in derogative detail, this is Eliot's first full depiction of an old man, and it will return more realised in 'Prufrock's Pervigilium', the subsequently unpublished lines in Eliot's notebook linked to 'The Love Song of J. Alfred Prufrock' (and considered in detail later with that poem). It is important to note that the Body of the title is represented deteriorating in old age, as opposed to the youthful Soul of the narrator. The old man is blind, which will be a recurrent characteristic of Eliot's old men, and both his respiration and mobility are failing; there was a manuscript deletion of "groans and mutters" which would have provided pain and potential dementia as additional symptoms of senescence.[33] The old man's surroundings serve to emphasise his lowly status. Instead of aiding either his eyesight or his mobility, his stick uselessly "pokes and prods", and does so at "withered leaves", themselves a symbol of both the passage of time and of physical decay. And he acts with "senile patience",[34] the consequence of an elderly absence of responsibilities and demands, encapsulating the purposelessness of geriatric life. As Jewel Spears Brooker observes, "The old man's offence is intruding himself into the view of the young man; simply by being a fellow human being and existing in the same space, he disturbs the composure of the young man".[35]

In the subsequent 'Bacchus and Ariadne: 2nd Debate between the Body and Soul' (1911) the Body is again depicted deteriorating with age, as "Time began again its slow / Attrition on a hard resistant face".[36] This attrition by time recalls the effacing by the years in his graduation 'Ode'. Thirty years later, in 'The Dry Salvages', Eliot will acknowledge that while the corporeal may be eroded by time, experiences, like others' torment, remain unworn by its attrition—and that, in the orbicular style of *Four Quartets*, time both destroys and preserves—but here, ageing is portrayed purely as a linear process of physical deterioration.

Several images of ageing in Eliot's juvenilia, like the blind old man in 'First Debate between the Body and Soul', will recur in his subsequent work. The slightly balding Prufrock is prefigured by "Life, a little bald and gray" in 'Spleen' (1910),[37] an image of life itself as gently ageing. The "aged sibyl in your eyes" in 'Goldfish (Essence of Summer Magazines)' (1910)[38] prefigures the epigraph to *The Waste Land*; while "I feel / As if I'd been a long time dead", in 'The Little Passion' (1911),[39] describes a detachment from the process of ageing within life later seen in the early manuscript of *The Waste Land*, where it became "'It seems that I have been a long time dead'".[40] That Eliot later repurposed these images of ageing affirms their ongoing significance, and illustrates the emergence of images of old age and considerations of ageing in the nascent poet which is expanded upon in his later published work.

Perhaps the most significant of the early poems is 'The Death of Saint Narcissus', written somewhere between 1912 and 1915. Its first seven lines were recast in *The Waste Land*, which together with the earlier poem's themes of sensual self-absorption, androgyny, sacrifice and eventual salvation, have made it of critical interest. But it also contains Eliot's first depiction of an old man not just as an observation but as a character, with behaviour and attributes which both reflect Eliot's earlier writings and prefigure subsequent geriatric personae.

The central character in 'The Death of Saint Narcissus', seemingly an amalgam of Saint Sebastian and Narcissus, undergoes a series of transformations. In the third of these, he becomes a young girl, who is "caught in the woods by a drunken old man". The young girl and the old man metamorphose, leaving the character feeling "drunken and old".[41]

Unlike the involuntary physical deteriorations depicted in 'First Debate between the Body and Soul', this old man is twice described as "drunken". The *Oxford English Dictionary* distinguishes between 'drunk' and 'drunken'; the primary meaning of 'drunk' being a person "That has drunk intoxicating liquor to an extent which affects steady self-control; intoxicated, inebriated; overcome by alcoholic liquor".[42] But 'drunken', in "the more common current sense", means "Given to drink; habitually intemperate".[43] This is a behavioural rather than physical deficiency, which extends the negative characteristics of old age beyond the immediately discernible physical symptoms in Eliot's previous images. It reflects a lack of control, which Eliot in his critical writing decries in adulthood. (See Chapter 4).

The influence of his grandfather's temperance did not constrain Eliot's own youthful drinking, as evidenced in his 'Ballade of the Fox Dinner' (1909)[44] celebrating his undergraduate roistering; but it may have influenced a distinction between social drinking (as depicted in the draft opening to what became *The Waste Land*[45]), and the 'habitually intemperate', with his use of the latter as a deprecatory aspect of old age. Like blindness, drunkenness will recur in an old man in 'Prufrock's Pervigilium'.

The drunkenness of the old man in 'The Death of Saint Narcissus' is not presented as an explanation or an excuse for his sexual assault upon the young girl. Rather, it is part of the "horror" Eliot envisages in old age, and subsequently presented in the transformed Narcissus being "drunken and old". The young girl is "caught", underscoring her lack of consent, and the sexual assault illustrates aberrant behaviour as a further negative trait which Eliot associates with old age. There is also an emphasis upon age difference; it is not just a girl, but a young girl, who falls prey to the old man. This juxtaposition of old man, young girl and a sexual encounter will recur in 'Dans le Restaurant', where it engenders revulsion in the narrator. Stevenson's "decline of animal heat" in the elderly prefigures the cooling of sense which Gerontion will experience, and the expiring of sense in 'Little Gidding', but where the sexual urges of Eliot's characters do persist into old age, they lead to consistently unacceptable behaviour—sexual assault, unwelcome lewd reminiscence, and the voyeurism of Tiresias.

The outcome of the assault is that the old man knows "his own whiteness...his own smoothness". This version of this poem is printed in *Poems Written in Early Youth,* the jacket of which states that its collection was prepared "under the supervision of the author",[46] and is emended from the two manuscript drafts[47] in which the old man knows "*her* own whiteness...*her* own smoothness" (my emphasis). The consequence is that instead of experiencing the draft version's physical contrast of age and youth, and apprehending the distinction, the old man in Eliot's 'supervised' version achieves *self*-knowledge.

This allies the poem more closely with the myth of Narcissus, and with another old man, Tiresias. In Ovid's *Metamorphoses*, Liriope, the mother of Narcissus, consults Tiresias, the seer, and asks whether her son will live a long life. Tiresias pronounces that Narcissus will live to a ripe old age, "*si se non noverit*"—if he never knows himself.[48] In Eliot's poem the arrows of Saint Sebastian arrive after the sexual assault; age and youth become as one, the character achieves self-knowledge, and his body is surrendered. In addition, the prophetic ability of one of Eliot's most significant later old men has been displayed.

The quantity of references to old age in Eliot's youthful writing is remarkable in itself, but so is the repeated assertion by a young man that the prospect of old age is one to be anticipated with trepidation if not dread. Eliot's adoption of a *carpe diem* philosophy is not used to reflect the pleasures of youth, but rather an apprehension of the horrors of old age, together with a rejection of the values of his grandfather with which he had been raised. Geriatric characteristics of physical and mental deterioration, of aberrant behaviour and lowly status, which are established in this early work, will be repeated as negative traits and images in his subsequent work, deployed to examine questions of increasing intellectual sophistication.

'The Love Song of J. Alfred Prufrock' was Eliot's professional début as a poet, written in Eliot's notebook, *Inventions of the March Hare*, in July 1911, when Eliot was twenty-two, and first published in *Poetry* magazine in June 1915. The magazine's 'Notes' on contributors introduced the twenty-six year old "Mr T.S. Eliot" as a "young" American poet.[49] While many of Prufrock's characteristics have been considered in depth over the years, his age has received less attention. In a 1994 essay, Bruch Hayman cites five critics who regard the character variously and imprecisely as "of middling age", "middle-aged", "aging", of "unromantic middle-age", and, again, "middle-aged".[50] The poet C. Day Lewis was to write in 1934, "Prufrock is a middle-aged man — (it is generally accepted that Eliot as a poet was born middle-aged, and has not been getting any younger since)".[51] Elisabeth Schneider suggests that Prufrock is "a young man who has never been really young",[52] a description that might well be applied to Eliot himself; Robert Crawford ended *Young Eliot*, the first volume of his biography of the poet, with the conclusion that "it was as if he had never been young".[53] The aforementioned descriptions of Prufrock certainly present a character older than Eliot was when he wrote the poem. Indeed, Wyndham Lewis believed the poem to be "about some fictional character dreadfully troubled with old age". On first meeting Eliot, he therefore made no connection between the work and its clearly young author.[54]

Prufrock refers to his spot of baldness, a symptom of age, while his thin arms and legs suggest sarcopenia, the age-related loss of muscle mass and strength. He speaks of prior events he has seen and known, repeated experiences which suggest an increasing awareness of ageing and the passing of time. Moreover, Prufrock compares himself to Shakespeare's Polonius, a father of two adult children, described by Hamlet as a "tedious old fool".[55]

In 1962, aged seventy-three, Eliot gave an interview to the *Grantite Review*, the magazine of Grant's House of Westminster School in London, regarded as the oldest boarding house at a school in Britain. Sandwiched between items such as the awarding of colours and the next meeting of the Old Grantite Club, the interview with Eliot is not credited, and may, like other pieces in the magazine, have been contributed by a pupil. Published in a Question and Answer format, the relevant exchange in the interview is as follows:

> Q.: Prufrock. Did the person you had in mind represent the age, or is he a character from the "wasteland" ?
>
> T.S.E.: I could not have answered that question at all when I wrote Prufrock. It was partly a dramatic creation of a man of about 40 I should say, and partly an expression of feeling of my own through this dim imaginary figure. In the theatre, for instance, I always feel that dramatic characters who seem living creations have something of the author in them.[56]

It is particularly noticeable in this exchange that Eliot offered up first in his response the fact that Prufrock was "a man of about 40". Of all the characteristics of Prufrock which he might have revealed or expanded upon, Eliot chose to specify his age. Given the clumsy phrasing of the question, Eliot may understandably have thought that the questioner, when asking whether Prufrock was representative of "the age", meant 'the particular life stage', as opposed to 'the epoch'. Regardless of how he interpreted the question, it is nevertheless clear from his answer that to Eliot, Prufrock's specific age was a significant part of his dramatic creation, a projection into years almost twice those of his own at the time of writing. The "feeling of my own" which he expressed through Prufrock could have been any of the character's anxieties, but Eliot's particular concern about age is clearly one of them. When considering how 'old' an age of forty might have seemed to Eliot, it is worth noting that in 1911, when he wrote 'Prufrock', the average life expectancy of a US male was only 48.4 years.[57]

Eliot did indeed have a "feeling" of being old at the time of writing 'Prufrock'. In her memoir, his friend Brigit Patmore recalls:

> I remember very clearly one evening before August 1914, when Tom asked me to have dinner with him and talked for many minutes of how old he was. 'I warned you, Brigit,' his slow voice, lower and more impressive than usual. 'I am so old that it makes me despair.' He looked the same slender man as usual and not one day older than thirty — and he was probably younger than that.[58]

As an expression of Eliot's "feeling" when he composed the poem, of "something of the author" in the character, the line "I grow old…I grow old…" therefore seems particularly apposite. In 1960, Eliot wrote, "that line, if I remember rightly, was borrowed from Sir John Falstaff".[59] The line from Shakespeare's *Henry IV* is actually "I am old, I am old",[60] a more emphatic acceptance of the speaker's *present* age. Prufrock's combination of youth with apprehension of age is echoed in remarks reportedly made by Eliot in a seminar with students at the University of Chicago: "The young man in 'Prufrock' is meant to signify someone young and sportive and a man conscious of growing old".[61] To feel that one is growing old is not a common feeling in young men, who rarely perceive themselves as ageing on a continuum, but rather anticipate entering a distinct state of 'old age' at some future date. Eliot has projected his personal concerns about ageing into the character of Prufrock; whatever their actual lifestages might be, both Prufrock and Eliot regard themselves as presently growing old.

There is, however, a passage omitted from the published poem, in which Prufrock regards an 'old man', and thereby provides a vision of how later life might be. Titled 'Prufrock's Pervigilium', the passage concludes with the following lines:

> I fumbled to the window to experience the world
> And to hear my Madness singing, sitting on the kerbstone
> [A blind old drunken man who sings and mutters,
> With broken boot heels stained in many gutters]
> And as he sang the world began to fall apart…
> (The square brackets in the above quotation are in the original manuscript.)

The characteristics of this old man echo those of old men in Eliot's early unpublished verse, and will recur in later poems. The "old drunken man" recalls the "drunken old man" of 'The Death of Saint Narcissus',[62] together with the "blind old man who coughs and spits and sputters / Stumbling among the alleys and the gutters" in 'First Debate between the Body and Soul'.[63]

In the Pervigilium's old man, all three of the characteristics with which Eliot has previously described old men—age, blindness and drunkenness—are combined. It is as if Eliot possesses a single compound vision of old age, realised for the first time in these lines, which will recur subsequently. The order of the unpunctuated adjectives renders 'old' one of three equally pitiable characteristics—blind, old, and drunken. This is not a respected Elder, suffering blindness and drunkenness, which might be the case of a 'blind drunken old man'; but rather a man who is similarly blighted in vision, age and rationality. Eliot uses 'old' as a pejorative; age is an impairment, as much as blindness or drunkenness. The old man is looked down upon, both literally and metaphorically. "Sitting on the kerbstone" he would be physically beneath any passer-by, more so the narrator looking down from a window. His lowliness is delineated by his behaviour (sitting on the pavement, singing and muttering) and by his appearance (broken stained boot heels); and his characteristics are habitual, since the boot heels have been stained "in many gutters", and he is "drunken" rather than "drunk" (*cf*. the old man in 'The Death of Saint Narcissus'). Eliot has used the image of this old man to personify "my Madness"; madness not as opposed to sanity, but to rationality. It is "the world", not the individual, which "began to fall apart". Furthermore, in order to represent this kind of chaos, Eliot has drawn once again upon his image of old age. To Eliot, the antithesis of social order and an absence of control is represented by the appearance, behaviour and physical decline of an old man. The old age presented in the final passage of 'The Love Song of J. Alfred Prufrock' is therefore in clear contrast to that in 'Prufrock's Pervigilium'. When Prufrock declares that he "shall" roll up his trousers, possibly part his hair behind, and wear white flannels, his observations could be either expressions of determined choice, or of the inevitability of growing old; but the attention to appearance, the wearing of white, and the location on a beach as opposed to a gutter, all speak of a condition in old age well above that of the old man in Eliot's alternative vision.

This contrast only underlines the sadness anticipated in old age in the final section of 'The Love Song of J. Alfred Prufrock'. It is an old age of loneliness and isolation, of disappointment, of surrender, and of exclusion from a world of love and beauty. Once aware of the omitted 'Pervigilium' lines, it is impossible not to contrast the old man singing, which Prufrock hears, with the mermaids, whom he does not believe will sing to him. These may also have been inspired by the essay 'Crabbed Age and Youth', in which Stevenson writes, "to take a cue from the dolorous old naval ballad, we have heard the mer-maidens singing, and know that we shall never see dry land any more".[64] Despite losing hair from his head and weight from his arms and legs, Prufrock is not a distasteful figure in his ageing, like other old men in Eliot's poetry; his is a psychological rather than a physical decline, but no less pitiable for that. The poem, Gordon Symes writes, "is a study of middle age rather than old age, but a middle age informed throughout by premonitions of defeat, with which old age is synonymous in [Eliot's] earlier poetry".[65]

Notes

1. Eliot, T.S. 2015. *The Poems of T.S. Eliot Vols I and II*. Ed. Christopher Ricks and Jim McCue. Faber and Faber 1, 397 n120.
2. Eliot, T. S. 2022. *The Complete Prose of T. S. Eliot: The Critical Edition Vols 1-8*. Ed. Schuchard, Ronald, Brooker, Jewel Spears, Cuda, Anthony, Dickey, Frances, Formichelli, Jennifer, Harding, Jason, Javadi, Iman, Stayer, Jayme and Chinitz, David. The Johns Hopkins University Press and Faber & Faber Ltd. http://muse.jhu.edu/resource/1 2, 63.
3. Eliot, *Prose* 7, 792.
4. Eliot, Charlotte C. 1904. *William Greenleaf Eliot: Minister, Educator, Philanthropist*, Houghton, Mifflin.
5. Eliot, Charlotte C. 358.
6. Eliot, Jr., William G. 1882. *Lectures to Young Men*. American Unitarian Association.
7. Eliot, William Greenleaf, 'The Personal Faith of Jesus', excerpted in Holt III, Earl K. 2011. *William Greenleaf Eliot: Conservative Radical*, Second Edition, Village Publishers, 156–7.
8. Eliot, *Prose* 2, 63.
9. D'Ambrosio, Vinnie-Marie. 1989. *Eliot possessed: T.S. Eliot and FitzGerald's Rubáiyát*. New York University Press, 60.

10. Eliot, *Prose* 5, 632.
11. *The Rubáiyát of Omar Khayyám*. 1859. trans. Edward Fitzgerald. Macmillan, VII.
12. *Rubáiyát*, XXIII.
13. *Rubáiyát*, LXXIV.
14. Eliot, *Poems* 1, 227.
15. Ibid., 229.
16. Ibid., 230.
17. *Rubáiyát*, XXVIII.
18. *Rubáiyát*, XVI.
19. Eliot, *Poems* 1, 242.
20. Eliot, *Poems* 1, 236.
21. Crawford, Robert. 2015. *Young Eliot; From St Louis to The Waste Land*. Jonathan Cape, 136.
22. Eliot *Prose* 1, 8.
23. Eliot *Prose* 1, 589.
24. Stevenson, R.L.1881. 'Crabbed Age and Youth', *Virginibus Puerisque*. Thomas Nelson & Sons, 104.
25. Stevenson, 102.
26. Stevenson, 97.
27. T.S. Eliot. 1959. Interview by Hall, Donald, 'T.S. Eliot, The Art of Poetry No 1,' *The Paris Review*, 21, Spring–Summer 1959, 21–37.
28. Eliot, *Poems* 1, 223.
29. Ibid.
30. *Eliot, Poems* 1, 268.
31. Eliot, T.S. 1967. *Poems Written in Early Youth*. Faber and Faber.
32. Eliot, *Poems* 1, 239.
33. Eliot, T.S. 1996. *Inventions of the March Hare*: *Poems 1909–1917*. Ed. Christopher Ricks. Faber & Faber, 65.
34. Eliot, *Poems* 1, 239.
35. Brooker, Jewel Spears. 2018. *T.S. Eliot's Dialectical Imagination*. Johns Hopkins University Press, 21.
36. Eliot, *Poems* 1, 257.
37. Eliot, *Poems* 1, 239.
38. Eliot, *Poems* 1, 248.
39. Eliot, *Poems* 2, 578.
40. Eliot, T.S. 1971. *The Waste Land: A Facsimile and Transcript of the Original Drafts*. Ed. Valerie Eliot. Faber and Faber, 113.
41. Eliot, *Poems* 1, 270.

42. *Oxford English Dictionary*, "drunk (adj. & n.2)," December 2024, https://doi.org/10.1093/OED/1004759858.
43. *Oxford English Dictionary*, "drunken (adj.), sense 2," June 2025, https://doi.org/10.1093/OED/1112010919.
44. Eliot, *Poems* 1, 233.
45. ed. Valerie Eliot, 5.
46. Eliot, *Early Youth*, jacket flyleaf copy.
47. ed. Valerie Eliot, 93.
48. Ovid, *Metamorphoses*, III, 348.
49. Eliot, *Poems* 1, 373.
50. Bruch Hayman, 1994, 'How old is Prufrock? Does he want to get married?', *CLA Journal*, September 1994, Vol. 38, No. 1, 59–68.
51. Lewis, C. Day. 1934. *A Hope for Poetry*. Basil Blackwell, 18.
52. Schneider, Elisabeth. 1975. *T. S. Eliot: The Pattern in the Carpet*. University of California Press, 30.
53. Crawford, 424.
54. Lewis, Wyndham 1962, 'Early London Environment', in *T.S. Eliot: A Collection of Critical Essays*, ed. Hugh Kenner, Prentice-Hall, 29 (reprinted from *T. S. Eliot*, 1948, Ed. March and Tambimuttu, Editions Poetry).
55. Shakespeare, *Hamlet*, II, 2.
56. T.S. Eliot Interview, 1962, *Grantite Review*, Vol XXIV No. 3, 17.
57. 'Life expectancy in the USA, 1900-98', University of California, Berkeley: https://u.demog.berkeley.edu/~andrew/1918/figure2.html.
58. Patmore, Brigit. 1968. *My Friends When Young: The Memoirs of Brigit Patmore*. Ed. Derek Patmore, William Heinemann, 89.
59. Eliot, *Poems* 1, 396.
60. Shakespeare, *Henry IV* II, iv.
61. T.S. Eliot, Interview in *TIME* Magazine, 20, 13th November 1950, 53.
62. Eliot, *Poems* 1, 270.
63. Eliot, *Poems* 1, 239.
64. Stevenson, R.L. 1881. 'Crabbed Age and Youth', *Virginibus Puerisque*, Thomas Nelson & Sons, 103.
65. Symes, Gordon. 1951. 'T.S. Eliot and Old Age'. *The Fortnightly Review*. Vol CLXIX Jan-June 1951, 188.

CHAPTER 3

"I Grow Old, I Grow Old": 1916–1921

Abstract Eliot's early published poems contain a series of old men, in predominantly derogatory portrayals. They display both physical and mental decline, as well as personal neglect and aberrant behaviour. Characteristics of Eliot's old men from his juvenilia, such as blindness and drunkenness, now recur in his published work, and can be seen to have particular significance; while with 'Gerontion', his first titular old man, Eliot deepens his presentation of senescence, depicting psychological as well as physical aspects of old age, and deploying an elderly character in order to personify his concept of the historical sense.

Keywords Rhapsody on a Windy Night · Dans le Restaurant · Gerontion · FitzGerald

In the poetry which was published following Eliot's début, the theme of old age persisted. His first collection, *Prufrock and Other Observations*, contains three further references to old age in addition to its titular work; four, in a collection of twelve poems, mean that more than a quarter of its works, by a poet of twenty-eight, make reference to old age. In the prose-poem 'Hysteria' (1915), it is "An elderly waiter with trembling hands"[1] who tries to usher an embarrassingly hysterical lady and her companion, who narrates the scenario, outside into a restaurant's

P. Keers, *T.S. Eliot*, Modern and Contemporary Poetry and Poetics,
https://doi.org/10.1007/978-3-032-25050-6_3

garden. It is the waiter's embarrassment, and his concern at antisocial behaviour on his premises, which are reflected in his "trembling hands", rather than any physical characteristic of old age. His lack of control in the scenario reflects that of both the hysteric and the narrator, making the waiter an equal party in the situation. This is reinforced by the fact that, uniquely in Eliot's poetry, the waiter is described as "elderly", avoiding the pejorative manner in which Eliot invariably uses the adjective 'old' at this time. Implying a respect granted to 'elders', this choice of adjective suggests that unlike Eliot's other contemporaneous old men, the waiter is not being denigrated.

'Rhapsody on a Windy Night' was first published in *Blast 2* in 1915, but written in 1911,[2] and reflects a year which Eliot had spent in Paris. Echoing an OED definition of a rhapsody, now obsolete, as "A literary work consisting of miscellaneous or disconnected pieces; a written composition having no fixed form or plan",[3] the poem presents a disparate series of visual images of a desolate metropolis. Decay and dissolution is depicted throughout, often through erosion of the natural world by time, with references to dying flowers, smoothed driftwood, rust and dust, and a vivid image of an old crab:

> And a crab one afternoon in a pool,
> An old crab with barnacles on his back,
> Gripped the end of a stick which I held him.[4]

The significance of the crab's age is emphasised, by defining it as 'old' in an otherwise unnecessary elaborative diacope. The crab is male, described as "him" rather than 'it'. And the crab carries burdens of old age, the "barnacles on his back" which life has imposed, which also render his appearance visually unattractive. Eliot is drawing upon the same characteristics of old age which he has used in his contemporaneous descriptions of old men. There is an auditory resonance with the title of the Stevenson essay, 'Crabbed Age and Youth'; the etymology of 'crabbed' refers "to the crooked or wayward gait of the crustacean, and the contradictory, perverse, and fractious disposition which this expressed",[5] a disposition commonly associated with old men. Yet there is something of a reversal of the anticipated balance between age and youth, since the young narrator of the poem merely "held" a stick, while the old crab "gripped" it. To 'grip' suggests greater strength in the more agèd of the two, an attribute which Eliot does not commonly grant to his older men; but given that 'old' is once again a pejorative in these lines, that strength may be seen as exerted in desperation rather than determination.

During the subsequent years of 1917 and 1918, Eliot wrote several poems in French. "I hadn't written anything for some time and was rather desperate", he said in his 1959 interview for the *Paris Review*. "I started writing a few things in French and found I *could* at that period. I think it was that when I was writing in French I didn't take the poems so seriously".[6] Perhaps writing poems in French, and not taking them so seriously, also removed Eliot's usual control over his self-expression, because in 'Dans le Restaurant' (1918) we can observe him again using a repellent old man as a central character, but revealing a significant episode from his own past.

The poem is narrated by a diner in a restaurant. A waiter relates to the diner the story of an encounter he had when he was seven, with a little girl drenched in a rainstorm. He gave the girl primroses, tickled her to make her laugh and felt an erotic sense of power and frenzy, before the encounter was interrupted, to the waiter's regret, by a dog. The diner is disgusted by both the waiter and his story—but implies that he, too, once had a similar experience, which may be the cause of his reaction. There is evidence to suggest that Eliot himself had just such a childhood experience. In a letter of 1939 his colleague John Hayward described to Eliot two sexual episodes in his own childhood. Eliot recalled in reply his "first love affair at [...] the age of five, with a young lady of three, at a seaside hotel". He insisted that unlike Hayward's more erotic experiences, "My feeling towards her was expressed entirely by bullying, teasing, and making her fetch and carry". Yet he wrote that "That was my preparation for reading (some years later) the *Vita Nuova*".[7]

In *Vita Nuova*, Dante states that Beatrice first appeared to him when the author was nine. And when Eliot wrote *Dante* in 1929, he stated that "the type of sexual experience which Dante describes as occurring to him at the age of nine years is by no means impossible or unique. My only doubt (in which I found myself confirmed by a distinguished psychologist) is whether it could have taken place so late in life as the age of nine years. The psychologist agreed with me that it is more likely to occur at about five or six years of age".[8] Not only does Eliot take the age back to that at which he had his own "first love affair"; it is interesting that Eliot describes Dante's encounter with Beatrice as a "sexual" experience, something few other scholars have suggested.

It becomes clear in 'Dans le Restaurant' that the waiter's childhood encounter echoes one which the diner himself had experienced. "*De quel droit payes-tu des expériences comme moi?*", the diner asks—with what right do you pay for, or what entitles you to, experiences like mine? The poem presents a dialectic between the diner's memory of his own childhood experience and the repellent waiter's story. As George Whiteside summarises it, in one of the few studies of the poem, "Clearly, the diner has summoned disgust for the waiter's appearance to counteract the effect of the waiter's story. Disgust counteracts desire".[9]

In order to personify that disgust, Eliot turns once again to the character of an old man. This waiter is Eliot's first evidently 'old man' to appear in print and, contributing to the dialectic of the poem, his appearance is even more unsavoury than Eliot's previous unpublished descriptions of old men. He is described as "*delabré*", dilapidated; he stoops over the diner's shoulder, and has a wrinkled face. The diner observes that he was a "*vieux lubrique*", an old lecher, even at the time of the incident. Eliot's description of an old man is again associated with unpleasant physical characteristics. The waiter is depicted as garrulous, fat-bottomed and drooling (*baveux*). His clothing, in this case his waistcoat, is stained, and his hair is greasy. In the end, the diner offers the waiter money in order that he might use the public baths. He is, both literally and figuratively, a dirty old man.

The structure of the poem allows the narrator to express disgust at a sexual experience through the depiction of a repellent old man, perhaps even envisaging the waiter indulging in the incident in his current odious form rather than in his childhood. Yet the narrator is also careful to retain ownership of his own similar experience and, despite the reaction it engenders, baulks at the idea that the two may be comparable. The poem's contrast between youth and old age therefore also contrasts beauty and old age, echoing the juxtapositions presented in 'The Death of St. Narcissus'. In addition, the poem establishes that the experiences of youth may survive and retain their allure into a decrepit old age.

'Dans le Restaurant' portrays an old man who, consistent with Eliot's previous descriptions, is distasteful, and by whom the narrator is repelled. As previously highlighted, this same device is used by Eliot on a number of occasions, including utilising several of the same terms in order to personify notions ranging from irrationality to sexual distaste. It is a trope to which Eliot returns when representing the unacceptable, and to express rejection, revulsion or disdain.

The poem concludes with seven lines about Phlebas the Phoenician. There is little to tie them into the preceding poem, except that the Phoenician is described, as he will be in *The Waste Land*, as having once been handsome and tall. This could be seen as a deliberate contrast to the dilapidated and stooped waiter; or as a warning of a potential fate awaiting the young diner; or as an authorial observation, after another conflict in Eliot's poetry between youth and age, that all such conflicts end ultimately in death. However, the fact that Eliot was to detach the lines and reuse them almost verbatim in English as Section IV of *The Waste Land* suggests that he did not regard them as having exhausted their significance here.

Prufrock and Other Observations was published in the UK in 1917, but the collection was not published in the US. Eliot was desperate to demonstrate literary achievement to his parents in the States; writing to his US lawyer and patron John Quinn he explained that US publication of a book would help satisfy his parents that he had not "made a mess of my life, as they are inclined to believe".[10] Sadly, his father was to die the next day, in January 1919, aged seventy-five. Six weeks later, at the end of February 1919, Eliot suffered what he described to his brother as 'a sort of collapse".[11] In order to embody his loss, decline, and concern about failure, Eliot would turn once again to the trope of an old man.

'Gerontion' (1920) was Eliot's first dramatic monologue employing the persona of an old man. The title, a term used by Aristophanes and Xenophon, is derived from the Greek *geron* ('old man'), with a diminutive which renders it 'little old man'. Crawford records that the poem was "typed out not so long after his Greek-loving father's interment",[12] which may partly explain Eliot's use of a Greek title. The title also has a number of historic and literary resonances which are worth exploring in further detail. Gerontius, a Roman general, was responsible for abandoning his garrison at Thermopylae, when it came under attack from barbarians. According to the Greek historian Zosimus, Alaric, leader of the barbarians, sent messages warning of his approach. This news was no sooner communicated to Gerontius than he and the garrison retired and left the barbarians a free passage into Greece. This inaction is echoed by Eliot's Gerontion, who declares that he was "neither at the hot gates…",[13] 'hot gates' being a literal translation of Thermopylae.

The same allusion to Gerontius occurs in an 1865 poem by Catholic convert Cardinal John Henry Newman, later set to choral music by Elgar. In the poem, 'The Dream of Gerontius', the soul of a dying man, guided by an angel and taunted by demons, passes from his body to Purgatory and Heaven. The age of Gerontius is not specified, but his name clearly also derives from *geron*, and the poem describes a span of a life completed "from youth to eld":

> From youth to eld, from sire to son,
>   He lived, and toil'd, and died.
> He dreed his penance age by age;
>   And step by step began
> Slowly to doff his savage garb,
>   And be again a man.[14]

Of relevance here is the fact that Shakespeare also makes use of the archaic term 'eld' in *Measure for Measure*, where the Duke declares that an aged man "doth beg the alms of palsied eld".[15] This line immediately follows those from *Measure for Measure* quoted by Eliot in 'Gerontion's epigraph. It is also worth noting that Crawford records that "in Tom's sixteenth year" a church concert in St Louis included the Angels' Chorus from Elgar's work, "and given that this performance was one of the highlights of his church's year, Tom was bound to have heard about it. 'Gerontius' gave him a name that, like other names garnered from St Louis, lingered in his mind".[16] Equally, the name Geronte was used by Moliere, a dramatist with whose work Eliot was familiar,[17] for characters in two of his plays.

The loose leaves of the typescript of 'Gerontion' were inserted into Eliot's notebook, *Inventions of the March Hare*, and that emended typescript shows Eliot's original typed title 'Gerousia', subsequently scored through several times, with 'Gerontion' in handwritten block capitals beside it.[18] The Gerousia was the Council of Elders in the ancient Greek city-state of Sparta, part of Sparta's legislative and judicial structure, composed of twenty-eight members over sixty years of age (plus the state's two Kings). Sixty was the threshold age for Spartan military service. Membership was to be respected, "for the office of elder is a reward of virtue", wrote Aristotle, and "the elders are good men and well trained in manly virtue". At the same time, the philosopher questioned the wisdom of conferring such powers upon older men, "for the mind grows old as well as the body".[19]

The idiosyncracy of the word 'Gerontion' has facilitated the title being commonly regarded as the name of the narrator, and not just the title of the poem. The title 'Gerousia' would have made the narrator a representative of that class, rather than an isolated individual. 'Gerousia' would have suggested that the narrator's incapacities were not just his own, but were to be found within the cohort of respected Elders, both historic and contemporary. Gerontion may be seen by some as a "generic old man",[20] but there was clearly an intention for the character to be viewed as an individual, in the context of elders (*gerontes*) one is expected to respect, yet with the diminutive which renders him 'little old man' immediately deprecating the narrator.

As alluded to above, the epigraph used in 'Gerontion' is a quotation from Shakespeare's *Measure for Measure*:

> Thou hast nor youth nor age
> But as it were an after dinner sleep
> Dreaming of both.[21]

In her comprehensive study of 'The Epigraphs to the Poetry of T.S. Eliot', Jane Worthington opines that "The theme of 'Gerontion' is given in the first half line of the epigraph, whereas the tone and atmosphere of the poem are suggested in the following line and a half. Such completeness is rare in Eliot's epigraphs [...] The mind of Gerontion moves over history with just that broken, sudden kind of movement that the mind follows in a half-waking dream".[22]

However, Worthington neglects to consider the manner in which the epigraph relates specifically to Eliot's preoccupation with old age. In spite of Eliot's lamentation with feeling old ("so old that it makes me despair"[23]) he had nevertheless striven since his student days to present himself as 'mature' in a number of different guises, including personally, critically (see Chapter 4) and creatively. Yet at the age of thirty he had clearly not yet reached the physical 'maturity' of Gerontion. Both youth and age, therefore, were mental constructs to Eliot, not physical or chronological realities. Eliot *himself* "hast nor youth nor age" – the poet is "dreaming of both".

The epigraph also presents opposition to the very concept of a 'Gerousia', the unadopted title of the poem. If a man "hast nor youth nor age", then the experience and hence respect predicated to accrue with seniority cannot be assumed. Despite being an Elder, Gerontion, in David Rosen's summary, is "not ripe with age and the amassed wisdom of a long life but, rather, moribund, needing to be read to by an obvious and poor surrogate for the normal, developed past he lacks".[24] The automatic granting of respect and status on the basis of age is undermined.

Even before the poem itself begins, therefore, Eliot has expressed several concerns about old age. His title and epigraph speak of elders who may not merit the respect or display the maturity historically accorded to their age; of old age as a precursor to death; and of the possibility that characteristics of old age may be present in youth.

From employing old men as symbols and personifications in his earlier work, Eliot now adopts an old man as narrator. The poem opens with a declaration of his identity: "Here I am, an old man". It is a straightforward and emphatic declaration, which will be echoed three times. This repetition of his identity as "an old man" means that even characteristics which might have other causes, or occur at other life stages—"a dull head", "unnatural vices" or "a dry brain"—also become associated with his age.

His predecessors in Eliot's poetry were depicted primarily through outward display—soiled and damaged clothing, dirtiness, drunkenness and antisocial behaviour. Gerontion, however, reveals more sophisticated characteristics of age. He suffers from a loss of passion; from impairment of his other senses in addition to sight; and from loneliness. He is becoming frail; his hands are weak, his joints stiff. If his house is seen as a metaphor for his body, it is "decayed", and it is only "rented", foreshadowing death. It is also "draughty"; by the end of the poem it is clear that the wind is a metaphor for the coming of death, with the gull that flies against the wind in its concluding passage a representation of life's course in the face of death. Eliot will use this metaphor consistently in the future, with the elderly Simeon explicitly "waiting for the death wind",[25] until the metaphor is negated in 'A Dedication to My Wife' where, despite Eliot's old age, "No peevish winter wind shall chill"[26] his late-life passion.

Eliot's previous depictions of old age were predominantly physical and visual, but in 'Gerontion' he progresses into its psychological aspects. An essay in the journal *Neuropsychiatr*[27] confirms from a clinical standpoint that Gerontion accurately exhibits psychiatric symptoms of old age: "cognitive decline; a depressed, embittered mind-set with negative cognitions about the past and present; indications of nihilism; and several aspects of formal thought disorder. [...] Qualitative research on severe depression in late life suggests that Gerontion's experience in the poem mirrors patients' realities".[28]

It also mirrors Eliot's own state of mind at the time. As Crawford states:

> ...this new work would draw on worries about a wasted life, which his father's death intensified; it was fed, too, by failure to fight in war; by a sense of utter exhaustion; by trepidation about how the peace might turn out; by ongoing anxieties about sex, religion and death.[29]

The expression of these complex notions meant that the poem's depiction of an old man would have to go beyond the recurring characteristics of Eliot's previous representations; and as he had not witnessed any comparable decline in his father, he would have to find another inspiration for his portrayal.

The source of the poem's opening was acknowledged by Eliot to be from A.C. Benson's biography of Edward FitzGerald, the writer famous for his translation of *The Rubáiyát of Omar Khayyám* (which, as we have seen in Chapter 2, inspired Eliot in his youth). Indeed, passages of the two texts are virtually identical:

> Here I am, an old man in a dry month, / Being read to by a boy, waiting for rain. ('Gerontion', 1)
> Here he sits, in a dry month, old and blind, being read to by a country boy, longing for rain: (Benson, p142)

A key difference between them, however, is that, while FitzGerald is "longing" for rain, Gerontion is merely "waiting" for it, emotionless, passive and inactive. The fact that Gerontion is "being read to by a boy" immediately raises Eliot's frequent dialectic between age and youth, resonating with the suggestion in the poem's epigraph that neither actually exists. Having lost his senses, the elderly Gerontion's worldly contact now relies upon youth. FitzGerald refers in one of his letters to "reading, and being read to, part of my time",[30] so the incident which inspired the opening of 'Gerontion' was not an isolated occasion.

FitzGerald is described by Benson as "wearing blue glasses, and a shade over his eyes, which were often painful".[31] Failing eyesight is a recurrent aspect of old age in Eliot's descriptions; his old men, from the 'Pervigilium' to Tiresias, are frequently blind. Late in his life, the depth of this concern became clear. When Eliot in his sixties had fortunately avoided significant deterioration of his sight, he revealed that it was not simply the idea of the interpolation of a third party, a reader, between himself and literature, which troubled him—it was that the consequence of not being able to read would be an inability to write. In 'Some Thoughts on Braille', an address delivered in 1952, Eliot said:

> I like to hear poetry well read. But I depend still more on reading it to myself [...] when my first acquaintance with a poem is by hearing it read, I do not altogether take it in [...] If I were completely dependent upon readers I should never feel that I had attained complete intimacy with the poem itself, and with its author.
> That is why, if I were suddenly blinded, or if I found the world slowly dimming before my eyes, I should be thankful for the invention of braille. Whether at my age [64] I could master it, I do not know; but the hope of mastering it would sustain me. *For without being able to read, independent of others to read to me, I do not believe that I could write.*[32] [my emphasis]

This powerful confession in later life may explain why blindness was such a persistent and pitiable characteristic of the old men in his early poetry, for its prospect threatened Eliot's single lifetime ambition.

There are further similarities between the character of FitzGerald as portrayed by Benson, and that of Gerontion, which suggest that the biography, echoing many of Eliot's other concerns about ageing, may be a source for much more of the poem's depiction of old age than just its opening line. Benson's book begins with a surprisingly unenthusiastic description of the subject upon which he is embarking. "The life that it is here proposed to depict", he writes, was "a life singularly devoid of incident [...] it was a life not rich in results, nor fruitful in example".[33] Benson depicts a life of unfulfilled potential, containing only one significant achievement. "The *Omar* forms, as it were, a pedestal for his fame; without it, FitzGerald's other works would not have received, and, it may be frankly said, would hardly have deserved attention".[34] Perhaps reflecting Eliot's father's fears for his son, Gerontion in turn records his life by what it has *not* been: "I was neither [...] nor [...] nor".

Benson concludes by "estimating the character" of FitzGerald:

> On the debit side of the account stand a certain childishness of disposition, indolence, a weak sentimentality, a slackness of moral fibre, a deep-seated infirmity of purpose [...] We may regret the want of strenuousness, the over-developed sensibility which led him to live constantly in the pathos of the past, the pain of the contemplation of perishable sweetness.[35]

The person Benson describes is echoed in the character of Gerontion to such a degree that the two may be seen as complementary, and the former as a significant source and inspiration for the latter. Vinnie-Marie D'Ambrosio believes that "From the very opening of FitzGerald's biography Eliot would have become enmeshed with FitzGerald's 'eccentricity, listlessness, and melancholy'",[36] and that given Eliot's personal, marital and psychological situation at the time,

> The parallels to his own life that were to be found in a reading of Benson's FitzGerald would intensify his anxiety about his present condition, would arouse even more fear as a prophecy of what could lie in store for him in an old age like FitzGerald's.[37]

The traits which Benson delineates are a powerful contrast to the values of William Greenleaf Eliot with which Eliot had grown up. The character of FitzGerald, a man whose work the young poet much admired, directly contradicts the attitudes of Eliot's grandfather and father to work, responsibility and success, attitudes which had been imposed upon Eliot since childhood, and which he was concerned that his parents felt he had failed to achieve.

Significantly, Eliot recommended the *Letters of Edward FitzGerald* in the reading list for the extension lectures which he gave as a twenty-eight year old in 1916.[38] Benson also considers the letters of FitzGerald, and describes the personality they reveal as "irresolute, and listless",[39] both adjectives which can be applied to Gerontion. For Benson, the quality of FitzGerald's letters lies partly in the fact that "They preserve the moment, the hour, the scene; they indicate the thought just as it rose fresh in the author's mind".[40] Similarly, Gerontion's narrative establishes the moment from the outset, and his thoughts and memories spill forth abruptly as if fresh from his mind, perhaps suggesting that he has no time left to lose. In a letter of 1874, FitzGerald wrote: "I find life little worth now; not that I am unhappy, but so wofully [*sic*] indifferent".[41] Shortly afterwards, he wrote that "There is not much [worth] living for after seventy-four"[42]; and he was to die in his sleep aged seventy-five. The torpid mood of Gerontion echoes that of FitzGerald's last years.

Just as the traits which Benson identified in FitzGerald are antipodal to the philosophy of William Greenleaf Eliot, so we find the traits of Gerontion lambasted in the latter's sermons. In one of his *Lectures to Young Men*, Eliot's grandfather laments the situation of old men who have not followed a life of faith, but must recollect instead a life of dissipation, licentiousness, violation of trust and other sins. “What then”, he asks, “must be the feelings of him, who reads the records of a life, not only worthless, but wicked?”

> Think of one who, in the silent loneliness of old age, broods over recollections such as these! He feels that he does not comprehend the depth to which he has fallen; the light of eternity is needed to reveal that to him; but he knows enough to be covered with shame, and the despondency of his heart is but little better than despair. Young man! kneel down and pray to your God, that he may save you from such a close of life as this![43]

In late 1908, Eliot had read *The Symbolist Movement in Literature* by Arthur Symons, which introduced Eliot to Jules Laforgue and other modern French poets, and which he later stated had affected the course of his life.[44] There he encountered Symons' view that “we realise the infinite insignificance of action, its immense distance from the current of life”.[45] Eliot will return to this conflict in old age between action and contemplation, and directly challenge his grandfather's ethos, with ‘A Song for Simeon’. But it is a passage that Symons quotes from Maeterlinck which summarises the form of ‘Gerontion’:

> “I have come to believe,” [Maeterlinck] writes, in *Le Tragique Quotidien*, “that an old man seated in his armchair, waiting quietly under the lamplight, listening without knowing it to all the eternal laws which reign about his house, interpreting without understanding it all that there is in the silence of doors and windows, and in the little voice of light, enduring the presence of his soul and of his destiny [...] I have come to believe that this motionless old man lived really a more profound, human, and universal life than the lover who strangles his mistress, the captain who gains a victory, or the husband who ‘avenges his honour’.”[46]

‘Gerontion’ is Eliot's first significant work in which history, cultural references, and characters and images from post-War Europe, are all “whirled” into a contemporary consciousness. That consciousness has to combine an awareness of a past both personal and cultural; what Eliot describes in his essay ‘Tradition and the Individual Talent’, written in the same year as ‘Gerontion’, as “the historical sense”, containing “the whole of the literature of Europe from Homer”.[47] Gerontion's old age is required for his consciousness to present this historical sense, to be 'present' at both the "hot gates" of Thermopylae and the "warm rain" of the First World War. Grover Smith identified this aspect of Gerontion as Emerson's ambition for a man "to live all history in his own person".[48] It is a positive attribute of old age, which Eliot has discovered and will exploit with Tiresias in *The Waste Land*. In later life, Eliot would write of Shakespeare, Dante, Homer and Virgil that "the appreciation of their poetry is a lifetime's task, because at every stage of maturing – and that should be one's whole life – you are able to understand them better".[49] As a figure near the end of his "whole life", Gerontion occupies this space.

Its historical sense, its cultural references, its use of evocatively named characters, and the presence of themes such as aridity, all mean that 'Gerontion' can be seen as a precursor to *The Waste Land*. Gerontion himself embodies loss, failure and decline—both personal and cultural—and his old age (and rented, decaying house) can be regarded as metaphors for civilisation. Eliot himself rejected this interpretation; in a 1944 letter to Seán Ó Faoláin, he wrote that "'Old age' may of course be taken as 'the symbol of a declining civilisation', but I wasn't thinking about declining civilisations when I wrote it".[50] Yet in January 1922, after Ezra Pound's cuts to *The Waste Land* had rendered it too short for solo printing, Eliot had asked Pound "Do you advise printing Gerontion as prelude in book or pamphlet form?"[51] The use of the term "prelude' suggests much more of a sequential connection than might a term such as 'accompaniment'. But Pound replied, "I do *not* advise printing Gerontion as preface. One don't miss it AT all as the thing now stands. To be more lucid still, let me say that I advise you NOT to print Gerontion as prelude".[52]

In that same letter to Seán Ó Faoláin, Eliot wrote of 'Gerontion' that "it can only be said to have been the expression of a mood, its variations and associated or evoked memories". Amounting as that does to a confession of personal feelings and memories as sources of the poem, contradicting Eliot's *credo* of impersonality in poetry, it underscores the significance of Eliot embodying his own mood and memories in the monologue of an old man.

Notes

1. Eliot, T.S. 2015. *The Poems of T.S. Eliot Vols I and II*. Ed. Christopher Ricks and Jim McCue, Faber and Faber, 1, 26.
2. Eliot, *Poems* 1, 418.
3. Oxford English Dictionary, "rhapsody (n.), sense 2.a," December 2025, https://doi.org/10.1093/OED/9852398201.
4. Eliot, *Poems* 1, 18.
5. Oxford English Dictionary, "crabbed (adj.), Etymology," December 2024, https://doi.org/10.1093/OED/1118359142.
6. T.S. Eliot, 1959, interview by Hall, Donald, 'T.S. Eliot, The Art of Poetry No 1,' *The Paris Review*, 21, Spring–Summer 1959, 21–37.
7. Eliot, T.S. 2009–2025. *The Letters of T.S. Eliot Vols 1–10* Ed. Valerie Eliot, Hugh Haughton and John Haffenden. Faber & Faber, 9, 368.
8. Eliot, T. S. 2022. *The Complete Prose of T. S. Eliot: The Critical Edition Vols 1–8*. Ed. Schuchard, Ronald, Brooker, Jewel Spears, Cuda, Anthony, Dickey, Frances, Formichelli, Jennifer, Harding, Jason, Javadi, Iman, Stayer, Jayme and Chinitz, David. The Johns Hopkins University Press and Faber & Faber Ltd. http://muse.jhu.edu/resource/1, 3, 731.
9. Whitehead, George. 1976. 'T.S. Eliot's "Dans le Restaurant"', *American Imago*, Summer 1976, Vol. 33, No. 2, 157.
10. Eliot, *Letters* 1, 315.
11. Eliot, *Letters* 1, 323.
12. Crawford, Robert. 2015. *Young Eliot*. Jonathan Cape, 314.
13. Eliot, *Poems*, 1, 31.
14. Newman, John Henry. 1914. *The Dream of Gerontius and other poems*. Oxford University Press, 26.
15. Shakespeare. *Measure for Measure*, III, 1.
16. Crawford, 68.
17. Eliot refers to Moliere in comparison with Jonson in his 1918 'Syllabus for a Tutorial Class in Modern English Literature', Eliot, *Prose* I, 757.
18. Eliot, T.S. 2021. *The Gloucester Notebook*. ed. Robert Hyde, Galileo Publishers, 166.
19. Aristotle. 1885. *The Politics of Aristotle*. Trans Benjamin Jowett, Oxford, 55.

20. Schneider, Elisabeth. 1975. *T. S. Eliot: The Pattern in the Carpet*. University of California Press, 51.
21. Shakespeare, *Measure for Measure*, III, 1.
22. Worthington, Jane. 1949. 'The Epigraphs to the Poetry of T.S. Eliot', *American Literature*, March 1949, Vol. 21, No. 1, 1–17.
23. Patmore, Brigit. 1968. *My Friends When Young: The Memoirs of Brigit Patmore*. Ed. Derek Patmore,William Heinemann, 89.
24. Rosen, David. 2003. 'T. S. Eliot and the Lost Youth of Modern Poetry', *Modern Language Quarterly* 64:4, 473-494 https://doi.org/10.1215/00267929-64-4-473.
25. Eliot, *Poems* 1, 103.
26. Eliot, *Poems* 1, 219.
27. Wigand, M.E., Wiegand, H.F., Jäger, M. et al. 2017. 'Old man—young man: T.S. Eliot's *Gerontion* and the problem of identity', *Neuropsychiatr* 31, 32–38 https://doi.org/10.1007/s40211-017-0215-4.
28. Wigand, Wiegand, Jäger et al., 35.
29. Crawford, 314.
30. Benson, A.C. 1905. *Edward Fitzgerald*. Macmillan, 65.
31. Benson, *Fitzgerald*, 56.
32. Eliot, *Prose* 7, 753.
33. Benson, *Fitzgerald*, 1.
34. Benson, *Fitzgerald*, 136.
35. Benson, *Fitzgerald*, 200.
36. D'Ambrosio, Vinnie-Marie. 1989. *Eliot possessed: T.S. Eliot and FitzGerald's Rubáiyát*. New York University Press, 160.
37. D'Ambrosio, 159.
38. Eliot, *Prose 1*, 480.
39. Benson, *Fitzgerald*, 136.
40. Benson, *Fitzgerald*, 139.
41. Benson, *Fitzgerald*, 55.
42. Benson, *Fitzgerald*, 64.
43. Eliot, Jr, William G. 1882, *Lectures to Young Men*, American Unitarian Association, 24–27.
44. Eliot, *Prose* 4, 11.
45. Symons, Arthur. 1899. *The Symbolist Movement in Literature*. William Heinemann, 175.
46. Symons, 155/6.
47. Eliot, *Prose* 2, 105.

48. Smith, Grover. 1996. *T.S. Eliot and the Use of Memory,* Associated University Presses, 102; the line itself is from Emerson's essay, 'History', seventh paragraph.
49. Eliot, *Prose* 7, 484.
50. Eliot, *Poems* 1, 468.
51. Eliot, *Letters* 1, 629.
52. Eliot, *Letters* 1, 630.

CHAPTER 4

"Maturity of Mind"

Abstract From the outset, Eliot incorporates his attitudes towards age into his critical approach, as he seeks to position himself and his work away from adolescence and youth. Even as a young man himself, the description of another author as 'young' is sufficient to attract Eliot's criticism. He dismisses as 'adolescent' the poetry of Shelley and others, and he criticises poetic reminiscences of childhood as weak and indulgent. He creates an associated critical framework around concepts of 'maturity' and 'immaturity', with 'maturity' reserved for the most successful and acclaimed works by the likes of Dante, Shakespeare, Virgil—and, by implication, by the young Eliot himself. Only as he ages, and recognises the honesty and universal truths in the late poems of W.B. Yeats, does he acknowledge that maturity is a lifetime's pursuit.

Keywords Critical prose · Maturity · Classicism · Impersonality · Allusion · W.B. Yeats

Eliot's critical prose begins appearing in print coincident with the first publication of his poetry, and is similarly imbued with his *idée fixe* about age. He creates an associated critical framework around concepts of 'maturity' and 'immaturity', the latter becoming a term of scornful disdain, the former reserved for the most successful and acclaimed works

P. Keers, *T.S. Eliot*, Modern and Contemporary Poetry and Poetics,
https://doi.org/10.1007/978-3-032-25050-6_4

and, by implication, for those by Eliot himself. He published twenty-three signed and unsigned reviews in 1916, while the following year saw his first longer literary essays published in magazines like the *New Statesman*. His first collection of critical essays, *The Sacred Wood*, was published in 1920, and he would continue writing and publishing critical prose throughout his career.

From the outset, Eliot incorporates his attitudes towards age into his critical approach, as he seeks to position himself and his work away from adolescence and youth. This may partly have been intended to lend greater authority to his opinions. Although he refutes in his verse the assumption that wisdom accrues with age, he recognised that this widespread belief could affect the reception of the views of a literary critic in only his late twenties, and therefore wished to avoid the corollary assumption that youth equates to naivety. His adoption of an older persona may therefore be expedience; but it reflects the perspective observed in his early verse, together with a dismissal of youthful thinking which he expresses in a letter to his brother: "Young people", he wrote, "are incapable of any real understanding".[1]

In the epigraph to his first collection of critical essays, *The Sacred Wood*, Eliot juxtaposed a line from Byron's *Beppo* with a passage from the *Satyricon* by Petronius (from which he would also take the epigraph to *The Waste Land)*:

> *Intravit pinacothecam senex canus, exercitati vultus et qui videretur nescio quid magnum promittere, sed cultu non proinde speciosus, ut facile appareret eum ex hac nota litteratum esse, quos odisse divites solent... "ego" inquit "poeta sum et ut spero non humillimi spiritus, si modo coronis aliquid credendum est, quas etiam ad immeritos deferre gratia solet.*[2]

In this passage, Petronius records his meeting at an art gallery with a white-haired old man ("*senex canus*"), whose appearance seems to promise great things ("*magnum promittere*"). Introducing himself this old man, later named as Eumolpus, declares that he is a poet, and a poet of no mean talent, if you can trust the awards he has received; although, he admits, influence often sees awards conferred upon the mediocre.

Petronius's observation of potential in Eumolpus would seem appropriate to the ambition of an aspiring literary critic, while the false modesty of the poet's statement about his talent acknowledges the positive reception of Eliot's first published poems. Later in the *Satyricon* Eumolpus will pontificate about literature and declare that "the mind cannot conceive or be fruitful unless it is steeped in the mighty flood of literature",[3] a statement prefiguring *The Sacred Wood's* essay 'Tradition and the Individual Talent'. Potential, recognition, and the value of a historical sense are all concepts profoundly relevant to Eliot at this point in his career; the thirty-two year old Eliot associates with them through the adopted persona of a white-haired old man.

If Eliot was not still in his 'youth' he had only just left it behind; and while he had complained to Brigit Patmore of *feeling* old (Chapter 2) he would have been commonly regarded by the literary establishment of the time as a young writer. Ezra Pound described Eliot in a 1914 letter as "a young American",[4] and *Poetry* magazine had described him upon his début the following year as "a young American poet".[5] Nevertheless, in his reviews and essays he presents 'youth' as synonymous with creative immaturity, and sufficient cause in itself for critical dismissal. Reviewing *Thomas Hardy: A Study of the Wessex Novels* in 1916, the "conscientious conventionality of the table of contents" alone is enough to indicate to Eliot the "youth of the writer". Eliot was twenty-seven at the time; the author, H.C. Duffin, was twenty-five. His book is subsequently dismissed as "exuberant and immature".[6] Of *Resentment: Poems*, by Alec Waugh, Eliot writes: "Mr. Waugh is said to be very young, and to have written a novel. That is a bad beginning, but something might be made of him".[7] And in 1918, reviewing a book by E.B. Osborn of the lives of young men fallen during the Great War, Eliot opines that "We are a little wearied, in fact, by the solemnity with which Mr. Osborn accepts the youthful mind and the youthful point of view [...] Sympathy with youth is life; but acceptance of youth at its own valuation is sentiment; it is indifference to serious living".[8] Eliot clearly wished to distance himself from any association with 'youth'.

Drawing, as he admitted, upon his own experience, Eliot imbued puberty with great significance in the development of the poetic consciousness. In a 1932 Harvard lecture, he proposed that children up to the age of twelve or fourteen find some enjoyment in poetry, but often then abandon it. A small minority, however, find within it at that juncture an "almost overwhelming introduction to a new world of feeling",[9] and he recounts his own stimulating discovery of *The Rubáiyát of Omar Khayyám* at the age of fourteen (See Chapter 2). Writing later of Edgar Allen Poe, he is critical of work which he believes has failed to progress beyond puberty; he observes in Poe a "curiosity", something in which "a pre-adolescent mentality delights".[10] Eliot criticises Poe's lack of "coherence" and a "consistent view of life", and of lacking "coordination of emotions"—all aspects of control and order, the absence of which, Eliot concludes, identifies work by a man "whose emotional development has been in some respect arrested at an early age".[11] Although Eliot avoids providing what he describes as "any possible psychological or pathological explanation" for such work, he remarks upon Poe's use of dreams of "ladies lost, or ladies vanishing, before they can be embraced", establishing a link with the sexual frustrations of puberty, and acknowledging puberty's physical as well as intellectual definition. The term 'puberty' is employed dismissively in his criticism, as when he writes in *The New Criterion* of George Bernard Shaw that "at two points, if not more, in his long series of plays Mr. Shaw reveals himself as the artist whose development was checked at puberty".[12]

Eliot employed 'adolescent' in his critical vocabulary in a similar manner. It should be noted that, like his concept of 'old age', Eliot's idea of 'adolescence' was chronologically fluid. In 1944, he recalled his personal experience of France, and remarked that "it was exceptionally good luck for an adolescent to discover Paris in 1910"[13]—the 'adolescent' Eliot was then twenty-one. It is also important to differentiate between his critical deployment of the term 'adolescent' and the support, encouragement and constructive criticism which Eliot offered to promising young poets. The poet and translator Edouard Roditi would later recall that when he was "barely eighteen" he submitted what he described himself as his "somewhat immature poems" to Eliot. "Very kindly, he suggested corrections", he wrote, "and encouraged me to continue submitting my poetry to him for guidance".[14] Eliot subsequently published three of his poems in *The Criterion*.

In 1920, writing on 'Modern Tendencies in Poetry', Eliot established a use of 'adolescent' which not only mocked youthful naivety, but placed it counter to his own poetic philosophy:

> Take the adolescent poet when he begins to write. He does not begin with a scientific spirit or much knowledge of his subject. His impulse is ejaculatory and imitative. He desires to extrude from himself some pressing and restless feelings in which he has become interested, and the shape they assume is decided by one or two admirations.[15]

Referring to a lack of creative discipline, the use of the term "ejaculatory" to describe the adolescent poetic impulse creates an immediate association with the self-stimulation of adolescent masturbation. "Ejaculatory" writing also implies that characteristic derided in his portrayals of old men, a lack of control; 'adolescent' creativity is portrayed as crudely physical, not intellectual. The language emphasises Eliot's distaste for such uncontrolled expressions of feelings in poetry; they run counter to his own philosophy, expressed in 'Tradition and the Individual Talent', that poetry should not be a "turning loose" of emotion—another term, like "ejaculatory", suggesting an uncontrolled release.

Eliot distances himself and his work from the issues he associates with 'adolescent' poets. By referring to the adolescent poet in the third person, and in derogatory terms, he establishes by implication that he is not—or, at the least, no longer—an adolescent poet himself. His position is that of having arrived at post-adolescence, i.e. maturity, and therefore entitled to scorn those who have not. This is the approach deployed in, for example, his criticism of Shelley:

> The ideas of Shelley seem to me always to be ideas of adolescence – as there is every reason why they should be. And an enthusiasm for Shelley seems to me also to be an affair of adolescence: for most of us, Shelley has marked an intense period before maturity, but for how many does Shelley remain the companion of age?[16]

Other inspirations from Eliot's youth, presumably now seen as mere "admirations", are similarly consigned to immaturity. Swinburne, a poet Eliot included in a list of discoveries who had excited him in childhood along with Omar Khayyám,[17] is merely "an adolescent who pretended to be writing poetry with much meaning, though it was really meaningless".[18] The French poet Laforgue, seen by many as a major influence on Eliot's emerging poetry, "remains, for us, imprisoned within his adolescence".[19] Eliot criticises Aldous Huxley's novel *These Barren Leaves* for its "adolescent self-analysis",[20] linking an exploration of one's personality to immaturity. One might observe of Eliot, as Jayme Stayer does, that "The writer who keeps insisting on the distinction between adolescence and maturity is a writer who is anxious to believe that he has achieved the crossover securely".[21]

Eliot's demand that poetry transcend the author's personality, and therefore the past which formed that personality, means that evidence of childhood in other writers is regarded as a failing. He decries it in authors such as Blunden, Lamb and Vaughan:

> We all know the mood; and we can all, if we choose to relax to that extent, indulge in the luxury of reminiscence of childhood; but if we are at all mature and conscious, we refuse to indulge this weakness to the point of writing and poetizing about it; we know that it is something to be buried and done with, though the corpse will from time to time find its way up to the surface.[22]

With a world-weary voice Eliot draws his audience into his own implied experience; to tolerate such retrospection displays a lack of the rigour and discipline which his philosophy demands, and which by implication he has exerted upon his own work. As with his criticism of 'adolescent' poetry, control over such 'indulgent' self-expression is demanded. Although Eliot would later state that listening to anecdotes of one's own childhood was "tedious",[23] he acknowledges here a pleasure in childhood reminiscence, suggested by the terms "luxury" and "indulgence"; but to allow it into one's writing would be not only a lack of control, but evidence that the author was not "at all mature". Only the sinister concluding metaphor suggests the possibility of darker personal reasons for Eliot's vigilant efforts to suppress his youth; the metaphor recalls the sprouting corpse of *The Waste Land*.

By criticising the presence of childhood reminiscence and adolescent traits in others, Eliot implicitly claims maturity for himself and his work. Its antithesis, immaturity, is a censorious criticism. This is evidenced as early as his undergraduate essay of 1909 on 'The Defects of Kipling':

> After all, there is one fatal weakness penetrating and marring almost everything to which Kipling sets his hand; accounting for several of the minor blemishes; it is his restless and straining immaturity. Nothing is so pathetic in literature as the immaturity which the practised brain cannot shake off, nor the practised hand conceal.[24]

'Maturity' was to become a central criterion in Eliot's early published criticism. As David Goldie observes:

> A noticeable feature of Eliot's early contributions to the Athenaeum is the marked absence of a consistent approach towards the tradition. The key words to denote approval in his critical vocabulary are not 'history' and

> 'tradition' but 'maturity' and 'rhetoric'; the qualities he derides are 'immaturity' and 'sentimentality'. The poets of the war are immature, bearing the hallmarks of Victorian sentiment rather than the less vitiated rhetoric of the Elizabethans.[25] The Victorians have inherited a debilitated romanticism based on the 'immaturity of genius' of Keats and Shelley.[26]

So Browning seems to Eliot "oddly immature"[27] when considered alongside Marvell. A poetry collection by Eliot's lifelong friend Conrad Aiken displays "immaturity of feeling",[28] while the immaturity of Henry Adams is "marked".[29] Eliot even suffers a "persistent feeling that Goethe always remained somewhat immature".[30]

Eliot's deployment of 'maturity' as a critical judgement came to the fore during what is known today as the Classicism-Romanticism debate, which unfolded in the 1920s between Eliot, as editor of *The Criterion*, and John Middleton Murry, the editor of *The Adelphi*. Murry instigated the debate by publishing an essay, 'On Fear: And On Romanticism',[31] in the September 1923 launch issue of his periodical, in which he argued that the true English tradition was inherently romantic rather than classical. Eliot responded in *The Criterion* of the following month, with 'The Function of Criticism',[32] and the two continued for several years to debate the issue in essays and reviews. To a degree, the debate was a promotion of their respective recently founded journals rather than an attempt to resolve passionately held positions. Their disagreement, believes Jason Harding, was manufactured out of opportunity. "Classicism defined as the antithesis of romanticism was a matter of personal taste and rhetorical tactics, not prima facie a question of ineluctable substantive differences".[33] When Eliot famously describes himself in his Preface to his essay collection *For Lancelot Andrewes* (1928) as "classicist in literature, royalist in politics, and anglo-catholic in religion", he continues by saying that "I am quite aware that the first term is completely vague, and easily lends itself to clap-trap".[34] And L.R. Sharma argues that both continued the debate in order to market their publications and their rhetorical skills to the literary community:

> Eliot and Murry were not interested in theorising on either romanticism or classicism. They were more interested in their own future (editorial and author-oriented) policies [...] These topics are focussed upon for reasons which could be best described as 'pragmatic' [...] this was an easily perceivable polarity between the two."[35]

Hence Eliot wrote to Lady Rothermere, patron of *The Criterion*, that he had written 'The Function of Criticism' for their journal "in the attempt to give the liveliness of controversy and the noise of battle".[36] But in arguing for Classicism, Eliot used and developed his notions of 'maturity':

> With Mr. Murry's formulation of Classicism and Romanticism I cannot agree; the difference seems to me rather the difference between the complete and the fragmentary, the adult and the immature, the orderly and the chaotic.[37]

This association of the adult with the orderly, the immature with the chaotic, continues the association Eliot has previously drawn between puberty and adolescence and an absence of control. His association of the adult and the orderly is a logical corollary. In a passage which was omitted from the version Eliot later published in his *Selected Essays*, he went further:

> the romantic is deficient or undeveloped in his ability to distinguish between fact and fancy, whereas the classicist, or adult mind, is thoroughly realist – without illusions, without day-dreams, without hope, without bitterness, and with an abundant resignation.[38]

Eliot will declare his definition of the Classic in *What Is a Classic?* (1944), which begins with the assertion that "If there is one word on which we can fix, which will suggest the maximum of what I mean by the term 'a classic', it is the word maturity".[39] Duncan Kennedy's analysis of that definition is apposite here:

> Eliot constructs a theory of cultures which is expressed in terms of the human body and its growth and decay. The *telos* or goal of such a culture is held to be 'maturity', a maturity of mind and a maturity of manners. Maturity of mind, equated with the use of language, is expressed by means of a contrast with youth and old age, and in terms which delicately evoke the sexual characteristics associated with each; maturity, it should be observed, comes to connote a timeless stasis between periods of life associated with those archetypal symbols of the passage of time, growth and decay.[40]

A problem therefore arose for Eliot with a conflict between his notion of 'maturity', wedded to the human body, and the role for the individual which he expounded in 'Tradition and the Individual Talent', with its "impersonal theory of poetry". In his criticism of Henry Adams, Eliot opines that:

> It is probable that men ripen best through experiences which are at once sensuous and intellectual; certainly many men will admit that their keenest ideas have come to them with the quality of a sense-perception; and that their keenest sensuous experience has been "as if the body thought." There is nothing to indicate that Adams's senses either flowered or fruited.[41]

This corporal failure to "ripen", "flower" or "fruit"—all synonyms for physical maturity—is for Eliot "what one feels in him as immaturity, indeed as a lack of personality".[42] But 'Tradition and the Individual Talent' questions the value of such maturing personal experiences to poetry, and demands an absence of personality from the poet:

> the poet has, not a "personality" to express, but a particular medium, which is only a medium and not a personality, in which impressions and experiences combine in peculiar and unexpected ways. Impressions and experiences which are important for the man may take no place in the poetry, and those which become important in the poetry may play quite a negligible part in the man, the personality.[43]

It is the quality of performance as a medium, rather than as a personality, which indicates maturity as a poet:

> the mind of the mature poet differs from that of the immature one not precisely in any valuation of "personality," not being necessarily more interesting, or having "more to say," but rather by being a more finely perfected medium in which special, or very varied, feelings are at liberty to enter into new combinations.[44]

An adaptation of his definition of 'maturity' followed, as David Goldie remarks:

> Maturity is a term that too readily applies to personal qualities for Eliot's purposes. By hypostatizing the attributes of maturity into the substance of tradition, Eliot can mark the qualities he values with the stamp of impersonal authority; he can give the qualities he has formerly associated with literary maturity the semblance of an extra-personal sanction."[45]

The literary philosophy which Eliot crystallised in 'Tradition and The Individual Talent' required Eliot himself to write as a medium, not a personality, and so to suppress any evidence of a past, a youth and adolescence, which might have forced personal influences upon his consciousness. The result is summarised by David Rosen:

> The consequence of this famous "escape from personality" however, is [...] Eliot's strenuous insistence on his own maturity, the poet disallowing any *developmental* or historical understanding of self and so needing simply to *assert* with sustained vehemence his own adultness and knowledge.[46]

Eliot's personal claim to literary maturity is made primarily through a series of critical definitions which assert maturity for himself inferentially. "Maturing consists largely of the taking in and digesting various influences",[47] he writes. Eliot demonstrates in both his poetry and his prose the ingestion of various influences; *ergo*, Eliot has matured. He maintains that "Maturity of mind: this needs history, and the consciousness of history"[48]; this prevalent characteristic of his own work must therefore demonstrate his own maturity of mind. And in one of his most famous edicts the poet who observably 'steals' from other sources declares that "Immature poets imitate; mature poets steal".[49]

Kathleen Woodward suggests that "The gift of history and tradition is a partial answer to the anguish of old age and death", and that in such 'stolen' material "the past is recovered and placed in the present, itself renewed, a renewing force".[50] James Longenbach believes that "By employing allusions, "Eliot self-consciously made his poetry difficult, the property of a specialist, in order to increase the status of poets",[51] but that

> While Eliot's very practice of allusion was designed to enhance his status as a professional poet, his acknowledgments of them (Dante but not Whitman) were designed to emphasize the particular sense of tradition that underwrote his modern "pantheon of literature." In more ways than one, then, Eliot's allusions were part of a self-consciously political program. Whitman and Tennyson were crucial, Eliot knew, to his personal life as a poet; but they could not be part of his public mission to alter the terms of his literary culture.[52]

Again therefore we see, as observed earlier, "the poet disallowing any *developmental* or historical understanding of self".

In 1944, Eliot blurs the emphatic definitions of maturity which that early criticism contained. "To define maturity without assuming that the hearer already knows what it means, is almost impossible",[53] he writes.

> If we are properly mature, as well as educated persons, we can recognize maturity in a civilization and in a literature, as we do in the other human beings whom we encounter. To make the meaning of maturity really apprehensible – indeed, even to make it acceptable – to the immature, is perhaps impossible."[54]

Yet as Eliot aged, he was able to make a more direct association between maturity and the passage of time than in his early criticism (when, as we have seen, despite telling others that he "feels old", relatively little of his life had actually passed). In a 1932 lecture at Harvard, aged forty-four, he acknowledged the passing years when he said that "The perception of *why* Shakespeare, or Dante, or Sophocles holds the place he has is something which comes only very slowly in the course of living".[55] The following year, at Washington University, it is recorded that he began his lecture by stating of Shakespeare, "he is one of those very few poets whom it takes a whole lifetime to understand. One reads him every year with more understanding, for it takes increasing maturity to know him".[56] This could be seen as rendering questionable Eliot's assertion of his own achievement of maturity in his late twenties.

It is with his acknowledgement of the qualities of the late poems of W.B. Yeats that Eliot finally recognises and accepts the significance of chronological age in relation to poetic maturity. Eliot was profoundly critical of Yeats's early work. In a review of 1919, Eliot writes that Yeats is "not of this world", and has "a mind in some way independent of experience [...] in which perception of fact, and feeling and thinking are all a little different from ours".[57] Eliot accuses Yeats of dreaminess, incoherence, egoism and crudity.

In *After Strange Gods* (1934) Eliot chastises Yeats for attempting to create an individual religion, one which lacked spiritual significance, and for poetry which was trifling, eccentric and provincial. He was particularly concerned about Yeats's involvement with psychics and the supernatural, and his use of folklore and myth—or as Eliot described them, "gnomes, hobgoblins, and astral bodies".[58] In private, he could be even less judicious; in a Faber reader's report in January 1934, he began with: "Uncle William is looney as ever".[59] The animadversion was reciprocated. "I think of him as satirist rather than poet", Yeats declared, writing about Eliot in his introduction to the *Oxford Book of Modern Verse* (1936). He described Eliot's verse as "grey, cold, dry", exhibiting "unexaggerated plainness", "rhythmical flatness" and "monotony of accent".[60] The two pursued what Richard Ellmann described as a "long, languid incompatability".[61]

In a lecture on 'The Modern Mind' in 1933, Eliot accuses Yeats of trying to achieve in his verse something akin to drug-induced exaltation, and again recites a list of the symbolic "paraphernalia" which he observes populating Yeats's early poetry. But then Eliot declares that:

> by a great triumph of development, Mr. Yeats began to write and is still writing some of the most beautiful poetry in the language, some of the clearest, simplest, most direct.[62]

Eliot returns to this "development" in subsequent essays and lectures to explain his apparent critical *volte face*. Marking Yeats's seventieth birthday in *The Criterion* in 1935, he declares that this "development" by the Irish poet "is not merely genius, it is character",[63] and that it provides a role model for younger poets, as Eliot can think of no other poet, even among the greatest, whose development has been over a longer period. This marks an unusual acknowledgement by Eliot of the benefits of longevity. Following his death, Eliot lectures on Yeats at the Abbey Theatre, Dublin, and expands further upon Yeats's development which, he believes, "is not fully evinced until the volume of 1914, in the violent and terrible epistle dedicatory of *Responsibilities*". Eliot also praises as "very impressive" lines from Yeats's poem 'The Spur'.[64]

However, that 1914 collection, *Responsibilities*, was published *prior* to Eliot's damning critical remarks of 1919 and 1934. Either Eliot did not wish to acknowledge work which would temper his vociferous criticism, or he did not perceive the development which he was later to praise so fulsomely. Eliot did reappraise some of his critical opinions as he aged; his shift between 'Milton I' (1936) and 'Milton II' (1960) being the most notorious. But it is perhaps telling that Eliot begins to praise Yeats's later poetry only as Eliot himself reaches the age at which Yeats composed it.

In the opening 'epistle dedicatory' of *Responsibilities* to which Eliot refers, Yeats writes that "I have come close on forty-nine",[65] a declaration of his age which Eliot highlights in his Dublin lecture. Eliot delivered his 1933 lecture at the age of forty-four, and his Dublin lecture on Yeats at fifty-one. Eliot regards himself and Yeats as contemporaries, a term he uses in both his 1933 lecture notes[66] and in his Dublin lecture,[67] although Yeats was actually twenty years older, and the *rapprochement* seems coincident with Eliot attaining the age at which Yeats wrote his later works. Of Yeats's poem 'The Spur', Eliot asks in 1940, "To what honest man, old enough, can these sentiments be entirely alien?"[68] With the phrase "old enough" Eliot acknowledges that he could only appreciate such sentiments as he reached a similar age himself.

Development was an attribute Eliot was eager to achieve himself as he aged. In his 1940 lecture, Eliot declared:

> That a poet should develop at all, that he should find something new to say, and say it equally well, in middle age, has always something miraculous about it.[69]

He was critical of poets like Wordsworth who "went droning on… to the verge of the grave"[70] and had always been concerned about repeating himself; submitting an early draft of 'East Coker' to John Hayward he is concerned that "it may be quite worthless, because most of it looks to me like an imitation of myself".[71]

Yeats provides Eliot with an example of how emotions experienced in old age, which Eliot now acknowledges, could provide poetic material. A brief summary of the treatment of old age in Yeats's late poems would incorporate their questioning of the value of past achievements and the purpose of continued striving. They acknowledge physical degradation, and personal responsibility, and portray the presence of emotions such as lust, rage, despair and humiliation, discussing and drawing upon all such negative characteristics as sources of poetic inspiration. Yeats writes primarily in the first person, eschewing personae. The style of Yeats's later poems reflects the more direct manner towards which Eliot was moving, and Eliot's praise in 1933 of the clear, simple and most direct poetry in late Yeats accords with the tone employed in passages of *Four Quartets*. Eliot identifies clarity, honesty and vigour in Yeats's poems; but it is the confessional candour of Yeats about his old age which has the most profound resonance with Eliot.

In the Dublin lecture, Eliot acknowledges "the honesty with oneself expressed in the poetry". Yeats's candour about old age, Eliot believes, is not simply an individual confession, but a potentially shocking revelation of a universality, the nature of "most other" old men. Those who do not accept this revelation "will refuse to believe that *they* are like that".[72] Eliot's emphasised third-person plural confirms his acknowledgement that he *is* like that. Yeats writes in the first person, at an age and status commensurate with Eliot's own, of bitterness, disillusionment and self-recrimination in senescence. In Eliot's critical *volte face* we see his recognition of this expressiveness, contradicting his earlier commitment to poetry as an escape *from* emotion and personality, as providing him with an example of a poet "who, out of intense and personal experience, is able to express a general truth; retaining all the particularity of his experience, to make of it a general symbol".[73]

In his recognition of late Yeats, Eliot reverts to the 'personal' quality of maturity which he abandoned after 'Tradition and the Individual Talent'. "Maturing as a poet" he writes, "means maturing as the whole man, experiencing new emotions appropriate to one's age, and with the same intensity as the emotions of youth".[74] It is perhaps an acknowledgement of experiences Eliot is encountering for himself as he ages; it is certainly a form of expression which Eliot is adopting in writing *Four Quartets*, and which will culminate in 'Little Gidding', drafts of which were written on the verso of notes for the 1940 Yeats lecture.[75]

In 1950, aged sixty-one, Eliot says in a talk on Dante that "One test of the great masters [...] is that the appreciation of their poetry is a lifetime's task, because at every stage of maturing – and that should be one's whole life – you are able to understand them better".[76] With his entitlement in increasing age to claim years of literary engagement behind him, the term 'lifetime' here looks forward as well as back, to yet greater appreciation and understanding. After years of claiming maturity, and criticising immature work, Eliot concluded in his old age that maturing is not a transition from adolescence, but a lifetime's pursuit.

Notes

1. Eliot, T.S. 2009–2025 *The Letters of T.S. Eliot Vols 1–10* Ed. Valerie Eliot, Hugh Haughton and John Haffenden. Faber & Faber, 1, 760.
2. Petronius. *Satyricon.* sec.83.
3. Petronius. *Satyricon.* sec 118.
4. Pound, Ezra. 1971. *The selected letters of Ezra Pound, 1907–1941.* New Directions, 40–41.

5. Eliot, T.S. 2015. *The Poems of T.S. Eliot Vols I and II*. Ed. Christopher Ricks and Jim McCue, Faber and Faber, 1, 373.
6. Eliot, T. S. 2022. *The Complete Prose of T. S. Eliot: The Critical Edition Vols 1-8*. Ed. Schuchard, Ronald, Brooker, Jewel Spears, Cuda, Anthony, Dickey, Frances, Formichelli, Jennifer, Harding, Jason, Javadi, Iman, Stayer, Jayme and Chinitz, David. The Johns Hopkins University Press and Faber & Faber Ltd, http://muse.jhu.edu/resource/1 1, 404.
7. Eliot, *Prose* 1, 734.
8. Eliot, *Prose* 2, 12.
9. Eliot, *Prose* 4, 591.
10. Eliot, *Prose* 7, 151.
11. Ibid., 152.
12. Eliot, *Prose* 2, 762.
13. Eliot, *Prose* 6, 514.
14. Roditi, Edouard. Letter, *The Jewish Quarterly*, no. 142 (38: 2, Summer 1991), 72.
15. Eliot, *Prose* 2, 213.
16. Eliot, *Prose* 4, 643.
17. Eliot, *Prose* 5, 632.
18. Eliot, *Prose* 4, 829.
19. Eliot, *Prose* 2, 661.
20. Eliot, *Prose* 3, 92.
21. Stayer, Jayme, 2013, 'T. S. Eliot as a Schoolboy: The Lockwood School, Smith Academy, and Milton Academy', *Twentieth-Century Literature* 59.4 Winter 2013, 622.
22. Eliot, *Prose* 3, 191.
23. Eliot, *Prose* 5, 430.
24. Eliot, *Prose* 1, 8.
25. Paraphrased from Eliot *Prose* 2, 10.
26. Goldie, David. 1998. *A Critical Difference; T.S. Eliot and John Middleton Murry in English Literary Criticism, 1919–1928*. Clarendon Press, 42.
27. Eliot, *Prose* 2, 319.
28. Eliot, *Prose* 2, 69.
29. Eliot, *Prose* 2, 44.
30. Eliot, *Prose* 4, 704.
31. Murry, John Middleton 1923. 'On Fear: And On Romanticism', *Adelphi* 1 (Sept 1923), 269–77.

32. Eliot, *Prose* 2, 458).
33. Harding, Jason. 2002. *The Criterion: Cultural Politics and Periodical Networks in Inter-War Britain*. Oxford University Press, 42.
34. Eliot, *Prose* 3, 513.
35. Sharma, L.R. 1994. *The T.S. Eliot-Middleton Murry Debate: The Shaping of Literary Theory: Modernist to Post-Structuralist*. Silver Birch, 16.
36. Eliot, *Letters* 2, 221.
37. Eliot, *Prose* 2, 460.
38. Eliot, *Prose* 2, 464, n24.
39. Eliot, *Prose* 6, 669.
40. Kennedy, Duncan. 1995. 'Tradition and Appropriation: T. S. Eliot and Virgil's "Aeneid"', *Hermathena*, Summer 1995, No. 158 (Summer 1995), pp. 73–94.
41. Eliot, *Prose* 2, 45.
42. Eliot, *Prose* 2, 44.
43. Eliot, *Prose* 2, 110.
44. Eliot, *Prose* 2, 108.
45. Goldie, 46–7.
46. Rosen, David. 2003. 'T. S. Eliot and the Lost Youth of Modern Poetry', *Modern Language Quarterly* 64:4, 473–494 https://doi.org/10.1215/00267929-64-4-473.
47. Eliot, *Prose* 5, 302.
48. Eliot, *Prose* 6, 676.
49. Eliot, *Prose* 2, 245.
50. Woodward, Kathleen M. 1980. *At Last, the Real Distinguished Thing: the late poems of Eliot, Pound, Stevens, and Williams*. Ohio University Press, 17.
51. Longenbach, James. 1994. '"Mature poets steal": Eliot's allusive practice', *The Cambridge Companion to T. S. Eliot*. ed. A. David Moody, Cambridge University Press, 178.
52. Longenbach, 180.
53. Eliot, *Prose* 6, 673.
54. Ibid.
55. Eliot, *Prose* 4, 592.
56. Eliot, *Prose* 4, 837.
57. Eliot, *Prose* 2, 72.
58. Eliot, *Prose* 1, 725.

59. Eliot, T.S. 2009-2025. *The Letters of T.S. Eliot Vols 1-10* Ed. Valerie Eliot, Hugh Haughton and John Haffenden. Faber & Faber,7, 59 n3.
60. *The Oxford Book of Modern Verse 1892–1935*. 1936. Chosen by W.B. Yeats, Oxford, xxi-xxii.
61. Ellmann, Richard. 1967. *Eminent Domain*. Oxford University Press, 89.
62. Eliot, *Prose* 4, 679.
63. Eliot, *Prose* 5, 260.
64. Eliot, *Prose* 6, 81.
65. Yeats, W.B. 1983. *The Poems*. ed. Richard J Finneran. Macmillan, 101.
66. Eliot, *Prose* 4, 776.
67. Eliot, *Prose* 6, 78.
68. Eliot, *Prose* 6, 84.
69. Eliot, *Prose* 6, 80.
70. Eliot, *Prose* 4, 626/7.
71. Eliot, *Letters* 9, 421/2.
72. Eliot, *Prose* 6, 83.
73. Eliot, *Prose* 6, 81.
74. Eliot, *Prose 6*, 78.
75. Gardner, Helen. 1978. *The Composition of Four Quartets*. Faber & Faber, 239.
76. Eliot *Prose* 7, 484.

CHAPTER 5

"I Tiresias, Old Man": 1922–1926

Abstract From its epigraph to its concluding character, old age suffuses *The Waste Land*. The multi-faceted nature of the poem provides an opportunity for the distribution of geriatric characteristics across multifarious characters and situations. Some elements—sexual transgression, failing eyesight, wrinkled appearance, an inability to act—have been encountered before in Eliot's poetry. Other aspects, including insomnia, anamnesis, deafness, voyeurism and kinesthesia, appear for the first time. His elderly characters can draw upon memory, which they possess in abundance (unlike the figures in the subsequent 'The Hollow Men') and they demonstrate with their old age life's relentless temporal linearity, in contrast to the non-linear kaleidoscopic form of the poem. The Sibyl, Marie, Phlebas, the pub ladies and Hieronymo all display particular traits of old age, deployed by Eliot for a range of purposes—and they all unite in Tiresias, "the most important personage in the poem", and Eliot's most significant blind old man.

Keywords The Waste Land · Sibyl · Marie · Phlebas · Hieronymo · Tiresias · The Hollow Men

P. Keers, *T.S. Eliot*, Modern and Contemporary Poetry and Poetics,
https://doi.org/10.1007/978-3-032-25050-6_5

Old age pervades *The Waste Land*. Both the Sibyl, the character who precedes the poem in its epigraph, and Hieronymo, the character who concludes the poem, are old. So are several of the characters who appear between them—along with Tiresias, the "personage"[1] Eliot tells us is the most important in the poem and unites them all. Given Eliot's persistent use of old age as a poetic device, and the depth and complexity which that device had provided him in 'Gerontion' (which was originally considered for inclusion as a 'prelude' to *The Waste Land*) it is not surprising that senescence features prominently in Eliot's iconic modernist poem. However, the varied ways in which Eliot depicts and utilises aspects of senescence, coupled with the multi-faceted nature of *The Waste Land*, provide an opportunity for the distribution of geriatric characteristics across multifarious characters and situations. Some of these characteristics—sexual transgression, failing eyesight, wrinkled appearance, an inability to act—have been encountered in his earlier poetry. Others, including insomnia, anamnesis, deafness, voyeurism and kinesthesia, appear for the first time. Some aspects are consonant across the elderly characters in *The Waste Land*, who are both historic and contemporary, "neither living nor dead", personifying the historical sense which the young in the poem lack. The elderly characters can draw upon memory, which they possess in abundance, as opposed to younger characters in the poem who personify immediacy and desire. And the elderly characters demonstrate with their age life's relentless temporal linearity, in contrast to the non-linear kaleidoscopic form of the poem.

The first geriatric character in *The Waste Land* appears before the poem has even begun. Eliot returns to the *Satyricon* by Petronius (which provided the epigraph, containing another elderly character, for his essay collection *The Sacred Wood* as discussed in Chapter 4) for the epigraph to *The Waste Land*:

> "Nam Sibyllam quidem Cumis ego ipse oculis meis vidi in ampulla pendere, et cum illi pueri dicerent: Σίβυλλα τί θέλεις; respondebat illa: άποθανεîν θέλω."[2]

The narrator in the *Satyricon*, a host at a dinner party, describes how he saw, with his own eyes, the legendary prophet the Sibyl of Cumae, hanging in a cage, and that when boys asked her "Sibyl, what do you want?" she replied, "I want to die". Russell Murphy writes that "The Sibyl was originally a single Greek oracle, an elderly woman who wrote her prophecies in ecstatic trances and then would toss to the seeker the leaves on which they were written. It would be up to the seeker, then, to make sense of the prophecy", presaging perhaps a reader interpreting the poem which follows. "The Cumaean Sibyl was a later but no less celebrated manifestation of this ancient Greek original".[3]

The Sibyl presents the most physically decrepit image of old age in Eliot's work—withered, shrunken and caged within life. We do not know her age at this point; Eliot's collaborator Ezra Pound suggests that she is "110 years of age, and some say she is over that".[4] She is not strictly immortal but, according to myth, condemned by Apollo to live and age for the number of years equivalent to the grains in a handful of sand. Once a significant personage, consulted for her prophecies, she now endures a life sentence of senescence, calculated by a handful of dust.

Whilst the Sibyl is the first of Eliot's elderly figures to openly ask for death, she will not be the last. Successors like Simeon and the Magus will link physical death to spiritual rebirth; but the Sibyl's desire is purely for a release from life, an escape from the physical burden of old age imposed upon her. It is a desire echoed by 'The Hollow Men' (1925), for whom "Life is very long".[5] Once, as a prophetess of divine significance, the Sibyl would have been consulted on major issues by senior figures; now, those asking the question of the Sibyl are children, establishing a tension between youth and age which will recur throughout *The Waste Land*.

The Sibyl also presents as an analogy for classical culture itself, with a declining status in the contemporaneous culture of *The Waste Land*. Both are of great age, an acknowledged presence remaining available for consultation, but are disregarded by the new generation, and treated without reverence or respect. The youth of *The Waste Land*, like those of Cumae, do not possess Eliot's 'historical sense'. Longevity is presented as a waning; instead of gaining stature—the narrative of successful ageing and the principle of the *gerousia*—both the Sibyl, and the classical culture from which she was born, have lost their status with age.

The poem itself could have begun with another repeat of Eliot's composite old man. In the opening passage of 'He Do the Police in Different Voices', the draft version of *The Waste Land*, we immediately encounter "old Tom, boiled to the eyes, blind".[6] Once again Eliot depicts an old man, drunk and blind. In this instance, however, his character is married, has both a circle of friends and a property to which he invites them, and is part of a roistering upper-class social circle ("Top hats and all") enjoying clubs and brothels. Providing champagne to his friends, "old Tom" echoes the drinking and clubbing of Eliot's Harvard days recorded in his 'Ballade of the Fox Dinner'. Intriguingly here, inebriation is a cause of blindness, not merely an accompanying characteristic. The character would have presented an antipode to the squalid old men portrayed previously, but was however excised along with this opening section. Consequently Marie is the first named character we encounter in *The Waste Land*.

Marie was identified by Eliot in a 1931 letter to Emily Hale[7] as Marie von Moritz, who had been living in the pension where Eliot stayed in Munich in 1911. He thought that she must have been about forty-two, but the life expectancy at birth in the German Empire, on the verge of the First World War, was just forty-nine,[8] so Eliot might well have regarded her as being in her final years. Forty-two was also the age at which Eliot published 'Ash-Wednesday', in which he would describe himself as "agèd".[9] Although Marie's senescence is not overtly specified, it can be inferred from the passage. Like many aristocratic and wealthier older Europeans in the early twentieth century, Marie travels to warmer climes in the South of the continent for the winter. "I would go south in the winter", the elderly Ivy says in *The Family Reunion*. "I would follow the sun, not wait for the sun to come here. I would go south in the winter, if I could afford it". Her sister Violet however dismisses those who do as "the vulgarest people [...] People with money from heaven knows where".[10] Marie's insomnia is also typical of an older generation. "Along with many physiologic changes seen with ageing, significant changes also occur in sleep and circadian rhythm across the lifespan".[11] The overall prevalence of insomnia symptoms in the elderly ranges from 30 to 48%, compared to 12% to 20% of the general population.[12]

Her familial link to the archduke and her politically charged declaration that she is a true German from Lithuania also suggest that Marie's life has connected with moments of historic significance. However, her reminiscence omits any such events in the memories to which it returns. Although she talked for an hour, the poem does not delineate this further. The weather is now more memorable to her than other, more significant events. The emotion of feeling free is of greater personal import than the geopolitics of freedom which surrounded her life. As her memory arcs back from old age to childhood, omitting the significance of her life in between, she illustrates the subjectivity of geriatric memory, an emotional version of personal history alternative to Gerontion's observations of the past.

Phlebas the Phoenician is introduced to us by Madame Sosostris as the drowned Phoenician Sailor before his appearance in Section IV of the poem. Her subsequent exhortation to fear death by water extends his personal fate more widely. The Sybil has already requested death, and suggested that far from being something to fear in old age, death is something to be desired; and death by water is to be portrayed as a rather gentle release from life, lulled by the current and erased from physical existence in "whispers".

(According to classical sources, Tiresias also suffered death by water in his old age, albeit of a rather different kind; he drank from the Tilphusa river, but "He was too old to withstand its cold", writes Athenaeus, "and died".[13])

Phlebas becomes detached in death from all that has defined his physical life as a sailor. He forgets the world of the senses, the acoustic and kinesthetic accompaniments to his long life. It is common knowledge that hearing frequently fails in the elderly, while the decline in old age of kinesthesia, the sense which provides understanding of the movement of the body, contributes to the propensity of falls in the elderly.

Phlebas has also forgotten the commercial activity which has defined his identity as one of the mercantile Phoenicians. The loss of identity in old age is a well-documented phenomenon. "That which we call our 'life'", writes Jean Améry, "the sum of what we have done and left undone, defines what we likewise still regarded as our life yesterday, that is, the years that are at best still left to us. Those we can already foresee as the homogeneous and monotonous repetition of our wasted time". The loss of the past and its accrued status, represents a loss of the future in terms of potential. "No one asks us any longer, 'What do you want to do?' All declare, dispassionately and unflinching, '*That* you've already done'".[14]

The crowd who flow over the bridge into the unreal City to work are similarly defined by their commercial activity. These are not youthful clerks and typists, but old men. In his 'Notes on The Waste Land' Eliot acknowledges Baudelaire as the source of his inspiration, but omits the title of the poem from which he draws his image: 'Les Sept Vieillards' (The Seven Old Men). With their memory still occupied with profit and loss, the sighs and blank stares of Eliot's crowd reflect not just the tedium of the single day ahead, but the foresufferance of a lifetime's work. Mr Eugenides escapes from his life of commerce into sensuality; Phlebas has forgotten both. His death is the release from working identity which those in the City seek.

Phlebas rose and fell metaphorically in life as he does physically in death; but, as he has forgotten the details of his working life, we lack Avery's "sum of what [has been] done and left undone", by which we judge the value of the elderly. If the noted similarity of imagery surrounding both Gerontion and Phlebas—the gulls, the seas, the characters whirled, and the ominous presence of the wind —urges us to compare their "sums", we sympathise less with Gerontion, who declares his sum of what in his life has been left undone, and more with Phlebas, whose sum of the done is unknown.

In a familiar trope of death, Phlebas's life passes before him—or, rather, he passes before his life. The only direction in which the whirlpool can take Phlebas at the end of his life is down; there is no hint of the spiritual ascension which Eliot's later old men may anticipate. His looks have deteriorated, as will happen to all who labour into age like him, but we are at least reminded that he was once attractive. And while he is now blind, the fate of so many of Eliot's old men, we have the image of his eyes replaced (as we are told twice elsewhere in *The Waste Land*) by pearls. "Look!", Madame Sosostris urges, in case we miss it. The haunting image is taken from Ariel's song in *The Tempest*, with Eliot finding in Shakespeare a beatific image to counter his own previous portrayals of blindness in old age as pitiable. When, at the end of *The Waste Land*, we are urged to sympathise, we will recall the tenderness of the portrayal of old Phlebas.

The final character taken from Western culture in *The Waste Land* is also its final old man. In Thomas Kyd's *The Spanish Tragedy*, Hieronymo is referred to five times, once by himself, as "olde Hieronimo" (Kyd's spelling), and the madness which afflicts him, following the murder of his son, is presented as possible senility: "Doubtless this man is passing lunatic, / Or imperfection of his age doth make him dote".[15]

In *The Waste Land*, Hieronymo offers to "fit", to provide what is needed or deserved, and may have done so already in the form of Eliot's preceding poem; but his credibility is dismissed, as the observations of the elderly often are, using an adaptation of the 1615 subtitle of the play: "Hieronimo is mad againe". As the concluding personage of *The Waste Land*, Eliot chooses an old man, one whose powerful memories are fundamental to his dramatic speeches in *The Spanish Tragedy*. Eliot thereby places the concept of memory as significant, both at the end of a long life, and at the end of the poem. But the perceptions of the elderly can easily be dismissed as insane, a response which can also be read as a foreshadowing of dismissive responses to *The Waste Land* and its author. Eliot consistently projects himself into the character of reminiscing old men, and "olde Hieronimo" tells us in *The Spanish Tragedy* that he too was a poet in his youth; which renders his lines following Eliot's quotation as an appropriate *envoi* to *The Waste Land*:

> Why then I'll fit you, say no more.
> When I was young, I gave my mind
> And plied myself to fruitless poetry:
> Which though it profit the professor naught,
> Yet is it passing pleasing to the world.[16]

By far the most significant old man in *The Waste Land* is Tiresias; and as the only two figures who can possibly *see* (Eliot's emphasis) "the substance of the poem",[17] Tiresias must be identified with Eliot himself. *The Waste Land* is not a dramatic monologue, presenting the narrative of a single personality. We are told in Eliot's 'Notes on *The Waste Land*' that its characters melt into each other and meet in Tiresias. Two personages each contain all of the poem's characters, as well as its allusions and references—Tiresias and, as *deus artifex*, Eliot himself. The two are synonymous. This is supported by Ezra Pound's comment on the manuscript of *The Waste Land*, when Eliot employs an indecisive "may". "You Tiresias", declares Pound in his marginal comment, "if you know know damn well or else you don't".[18]

Eliot will finally openly identify with Tiresias in the epigraph to his essay collection *After Strange Gods*. Tiresias, as we shall see, was fated to have his foresight ignored by the public. Fearing that his own perceptions might be similarly dismissed, Eliot addresses the audience for *After Strange Gods* through the collection's epigraph in the first person of Tiresias, speaking to Oedipus in Sophocles' *Oedipus Tyrannus*:

> καὶ ταῦτ' ἰὼν εἴσω λογίζου· κἂν λάβῃς ἐψευσμένον, φάσκειν ἔμ' ἤδη μαντικῇ μηδὲν φρονεῖν[19]

> Go inside and think on this – and if you find me mistaken, then say that I have no skill in prophecy.

With 'Gerontion', Eliot employed an old man to contain a consciousness broad enough to encompass all of Western civilisation. Again with

Tiresias he sought to illustrate in an elderly personage the historical sense, which he had defined in 'Tradition and the Individual Talent' as containing European literature from Homer to the present, and which Eliot claimed for himself in his critical writings, in his poetry, and by adopting in *The Waste Land* the perspective of the classical prophet.

The prophet addresses us directly three times in *The Waste Land* as "I Tiresias", and tells us boldly in two of those declarations that he is an "old man". This is not simply a character, described by Eliot as being an old man; this is a consciousness, repeatedly self-identified as that of an old man. This posits that the whole poem is viewed through the perspective of an old man. Yet when we look at the passage from Ovid to which Eliot guides us in his 'Notes on The Waste Land', there is no mention of Tiresias's age. We are told only that he is 'learned' ("*docti*"). Age is clearly significant to Eliot's Tiresias, as he tells us that he is old as frequently as he tells us that he has female breasts, his other fundamental physical characteristic. But there is nothing in the passage which Eliot cites (Ovid, *Metamorphoses* III, 320–338) to suggest that Tiresias is an old man. Eliot must therefore have derived Tiresias's old age from other sources, of which there are possibilities both in classical literature and in Eliot's nineteenth-century reading.

The passage from Ovid which Eliot quotes in his 'Notes' relates how Tiresias had once encountered two serpents mating, and upon striking them with his staff to separate them, had been transformed into a woman. Seven years later, encountering another pair of mating serpents, Tiresias struck those with the staff, and was transformed back "as he had been born". This experience in both sexes gave him the authority to arbitrate "a playful dispute" between Jove and Juno as to whether men or women took the greater pleasure from sex. Jove argued that women gained greater pleasure, and Tiresias took Jove's side—as a consequence of which he was condemned by Saturnia to eternal blindness. No god can undo what another god has done but, as the "playful dispute" did not warrant such a punishment, the Almighty Father granted Tiresias in compensation the power to know the future. There is no mention in this passage of Tiresias's old age. It is only in a later passage of *Metamorphoses* that Tiresias's age is evidenced, when he prophesies the fate of Pentheus, who mocks "the words of the old man", while Tiresias himself shakes his "hoary head".[20]

There is, however, an alternative classical story of Tiresias's blindness and prophetic powers, as related by Callimachus in his *Hymn V*, 'On the Bath of Pallas'. Callimachus tells us that whoever saw Athena naked, even unwittingly, "would see Argos for the last time".[21] A thirsty young Tiresias went to the fountain where his mother was bathing with Athena, and, unwittingly, Tiresias saw Athena naked, whereupon "night seized his eyes".[22] His mother wept and wailed, but Athena blamed "the laws of Zeus" and "the thread of the Fates"[23] which she could not undo. In compensation, however, she would make Tiresias "a seer", greater than any of the other prophets.

Among other gifts she would also grant him a very long "lifespan", confirming the old age which he declares in *The Waste Land*. Hesiod records that Tiresias would live as long as seven generations of humankind.[24] Furthermore, as a final gift, Callimachus records, "He alone, when he dies, will walk with understanding among the dead".[25] This foreshadows Tiresias's guidance of Odysseus through the Homeric underworld; his statement in *The Waste Land* that he has "walked among the lowest of the dead"; and the role which he will play among the denizens of the poem itself.

There are other classical sources for Tiresias's old age. In Sophocles' *Antigone*, Creon, King of Thebes, describes Tiresias several times as "aged" and "old".[26] Tiresias enters for his audience with the King led by a boy, a presence of old age and youth together echoing that in 'Gerontion'. Creon dismisses his prophecies, angrily accusing Tiresias of both dishonesty and avarice. This leads Tiresias to walk out, "so that [Creon] may vent his spleen against younger men".[27] It is a dignified exit, laying claim to the respect due to an older man. It displays both a disdain towards the behaviour of younger people, and an inability to take direct action, both of which foreshadow the behaviour of Tiresias in *The Waste Land*.

Neither Homer nor Dante, both of whom employ the character of Tiresias, describe him as old, but later sources may have influenced Eliot's depiction. Swinburne and Tennyson, both occurring extensively in Eliot's criticism, wrote dramatic monologues narrated by Tiresias. Grover Smith identified Swinburne's poem 'Tiresias' (1871) as a source, featuring as it does the construction "I, Tiresias".[28] A.V.C. Schmidt went further still, linking all that Tiresias in *The Waste Land* has "foresuffered" to Swinburne's description of Tiresias's prophetic skills as "with foresuffering bought".[29] In Swinburne's dramatic monologue Tiresias makes several references to his age. He is "full of blind years", "communes with the years", has had "centuries of high-thoughted life to live", and declares: "I am as Time's self in mine own wearied mind".[30]

Old age is similarly present in Tennyson's 'Tiresias' (1885). Eliot included Tennyson in his 1916 'Syllabus for a Tutorial Class in Modern English Literature', in which he recommended studying Tennyson's "Relations with his contemporaries"[31]; Tennyson's poem is framed with a paean and an elegy to Edward FitzGerald, the translator of *The Rubáiyát of Omar Khayyám* which so influenced the young Eliot, and whose biography provided the opening lines of 'Gerontion'. Eliot would recommend "the really great classical scholar"[32] exposed in two of Tennyson's poems, 'To Virgil' (1882) and "'Frater Ave atque Vale"' (1880), both of which were published in the *Tiresias* volume.

Tennyson's Tiresias is "winter-white"; his sight of Athena took place "long ago, / five-fold thy term of years". He believes his life has been a failure, cursed by "their unbelief, who heard / And heard not" his prophecies. "To cast wise words among the multitude / Was flinging fruit to lions".[33] This echoes Tiresias's lament in Euripides' *The Phoenician Women*, where he bemoans the role of a prophet, hated by men if he brings them bad news, and wronging the gods if he pities men and tells them lies. M.A.R. Habib observes: "This irony of simultaneous detachment and involvement is his perennial condition in literature".[34] It summarises his position in *The Waste Land*.

For much of Tennyson's poem, Tiresias argues passionately for the young Menoeceus to commit suicide, which he prophecies will save Thebes (as referenced in *The Waste Land*) from destruction. He acknowledges that this requires Menoeceus to believe him, unlike so many figures in his past, "...but thou art wise enough / Tho' young, to love thy wiser". The fate of Tiresias, as well as that of Menoeceus, depends upon the young believing the old, and acting accordingly. It is Eliot's familiar tension between youth and old age which will determine an outcome. But when his prophecy is believed, Tennyson's Tiresias echoes the wish of the Sibyl: "I would that I were gather'd to my rest". His own life *has* ended well, for his virtue, the saving of Thebes, has been achieved. The passivity of being "gather'd" into death as an old man contrasts with the suicidal action required of the youthful Menoeceus. It also foreshadows the inability to act, the separation of observer and participant, and the distancing from the actions of the young, which Tiresias displays in *The Waste Land*.

The poem's typist episode resonates most powerfully not with Eliot's cited passage from Ovid, but with the 'Bath of Pallas' story by Callimachus, with Tiresias both there and in *The Waste Land* a spectator observing intimate activity without permission. Observing the sexual activity of the young is a widely assumed trait of old men, and Tiresias also sees and erotically describes in detail the typist's underwear, before he watches the sexual act itself. His voyeurism echoes the aberrant sexual behaviour of other old men in Eliot's poetry such as Saint Narcissus, and the waiter in 'Dans le Restaurant'.

Eliot's citation of the 'Jove and Juno' story underscores Tiresias's intersex experience, which enables the prophet to embody and throb between both the male and female characters in *The Waste Land*. His intersex characteristics are most overtly expressed in his stated possession of female breasts; yet these are also indicators of his old age, for in both of his statements they are described as "wrinkled". This indicates breast atrophy, a natural effect of female ageing. Tiresias is old and so unable to suckle—or provide succour to—the young. It also suggests an association with the dessicated figure of that other aged female prophet from *The Waste Land*, the Sibyl.

In his second statement in *The Waste Land* Tiresias refers to his wrinkled breasts as "dugs". In the manuscript, the word provides a rhyme with a subsequent line which Pound excised for being "too easy": "Knowing the manner of these crawling bugs".[35] Without the justification of the rhyme, the term is striking and harshly animalistic. The earliest uses of the word cited in the *Oxford English Dictionary* apply to women, but later apply to animals, and the definition of 'dug' contained in the 1928 edition stated: "As applied to a woman's breast, now contemptuous",[36] a definition appropriate to Eliot's time of writing. As used by Tiresias the word not only mocks his geriatric inability to provide the young with the most basic nourishment, but also dissociates his aged breasts from any possible erotic quality.

Although ostensibly a "mere spectator"[37] Tiresias actually provides the perspective through which we ourselves perceive the typist episode. Blindness, which has previously served as a characteristic repeatedly appended to Eliot's old men alongside other physical failings, is a fundamental element in the Tiresias legends. His prophetic ability was granted in compensation for his loss of vision, metaphysical perception replacing physical sight, granting him the ability to "see" and for the events in the poem to be "perceived". The typist episode is not a Shakespearean scene with a third character, an old man, hiding 'behind the arras'. We see the typist's underwear because Tiresias sees it; we envisage the sexual encounter because Tiresias perceives it. It is his elderly perspective through which we view the young participants—bored, tired and indifferent on one side, bold, vain and patronising on the other—and pass judgement upon them. We are not the public who were so resistant to Tiresias's prophecies; we have the perspective of the prophet himself, and so know the truth of the vision. And his interjections remind us of the lifetime of experiences which underpin that vision; Tiresias knows that the clerk is "one of the low" because Tiresias has already "walked amongst the lowest of the dead". The weary ennui typical of an old man saturates his admission that he has already suffered everything that is taking place.

That Tiresias is "throbbing" may suggest that he can still gain sexual excitement from his voyeurism. We have seen, in 'The Death of Saint Narcissus' and 'Dans le Restaurant', that Eliot denegrates geriatric sexual feelings. Alternatively, Tiresias's detachment from the sexual activity of the youths, and his use of the verb 'suffer' to describe it, can suggest geriatric impotence, with his contemptuous attitude towards the clerk reflecting what John Richardson describes as "the aggrieved and anxious anger of a weak old man towards a vigorous, potent young one".[38] Potency and impotence, along with male and female, young and old, are two more of the lives between which Tiresias throbs. Ageing poses a threat to sexual activity in *The Waste Land*; just as the prophet's wrinkled breasts indicate a decline in arousal, so Lil is told that she "ought to be ashamed [...] to look so antique". Old age is something to be disguised, with smarter clothes and dentistry. Despite the barren sexual activities of the young within the poem, and despite the pub ladies' fear of unwanted children, life for these women revolves around the provision of "a good time", and once that passes in old age, their time will be called. Childbirth is seen, not in the context of new life but, with its risk of maternal mortality, as a threat to survival into old age; while Mrs Porter shows how motherhood can only condemn a young daughter to relive the squalid life of an aged parent, *ad infinitum*.

Given the tensions between youth and age which occur in the poem, and its publication shortly after the First World War, it is understandable that some perceived *The Waste Land* to be expressing to them what the influential critic I.A. Richards described in 1926 as "a clearer, fuller realisation of their plight, the plight of a whole generation, than they find elsewhere".[39] Years later, Eliot would reject this, dismissing the idea that he had expressed the disillusionment of a generation, and stating his specific dislike of the word 'generation'.[40] It is indeed hard to regard Eliot as representative of that post-War generation, particularly given the attitudes which he has displayed in his critical writing towards youth. Although a harsh description, it is perhaps easier to see Eliot at the time of writing *The Waste Land* as John Richardson describes him: "Someone in his early thirties who is already antique both in mind and (through sickness and sexual failure) in body".[41]

'The Hollow Men' (1925) is a poem of mood, an illustration of a state of despair which might be experienced at any age. Yet the condition of failure which pervades it imbues the line "Life is very long"[42] with a sense of suffering into old age. The greater the longevity, the greater the suffering, for their life contains no joy. The proximity to death and the end-of-life imagery in the poem are not confined to old age, but gain in intensity if the situation of the hollow men is approached through years of futility.

The hollow men share one of the primary characteristics of Eliot's old men—blindness. We are told that the hollow men are "sightless", and that eyes are failing in, or absent from, the land described. In 'Eyes that last I saw in tears' (1924), a minor poem which is "a preliminary sketch"[43] for 'The Hollow Men', the eyes are not only unable to produce tears, but are unable to be seen, a condition described as an "affliction".[44] The blindness of the hollow men can be alleviated, and the "eyes reappear", only through redemption, symbolised by the "multifoliate rose", an image which will eventually conclude 'Little Gidding'. As Dante crosses into Hell in the *Inferno*, those without hope of death are described as leading a "blind life"[45] Vision, both literal and spiritual, is absent in the hollow men.

Their "dried voices", immediately associating with the straw stuffing of the heads from which they emanate, can also be interpreted to reflect the dysphonia of age-related changes to the vocal cords. Their voices are, we are told, "As wind in dry grass", which echoes Gerontion's "dull head among windy spaces", and continues the association of wind with the end of life. The limp and insubstantial quality of a stuffed figure confirms the inability to act which Eliot frequently presents as a trait in old men. And the hollow men have, like Gerontion, lost their passion; their mouths have transitioned from sensual kiss to (futile) prayer. All suggest that the hollow men are agèd.

Absent from the hollow men are memories. The elderly characters in *The Waste Land* possess an abundance of memories, often expressed with a sense of loss, or with an understanding that their past has precipitated their present. 'The Hollow Men' is situated in the present and future. The absence of a past, of the experiences age, is one aspect of the characters' void, a memory loss typical of old age. The circumstances of the hollow men are consequently unrelieved in the poem by any memories which might either alleviate or justify their situation. "What might be said" writes Gordon Symes, "[…] is that the old men are hollow because they have been emptied – emptied of the secular affections, amongst other things, which still preoccupied the actors in Gerontion – and that they are now. conceivably, ready to be refilled".[46]

In a letter to his Italian translator in 1960, Eliot wrote that it was likely, when he said "stuffed men", that "the image of the Stravinsky ballet of Petrouchka was in my mind".[47] In Stravinsky's ballet, Petrushka is a puppet, stuffed with straw, brought to life by an old man, Charlatan. When Petrushka is 'murdered', Charlatan reminds the audience that Petrushka is just a straw-filled puppet, not a living character—only for the ghost of the slain Petrushka to appear, terrifying the old man. We shall see that this suggestion, that an old man is a controlling influence over the life of stuffed characters, has significance in the poem. Eliot concluded his remarks to the translator with "you hold the right stick to beat the old guy with",[48] perhaps echoing his poem's epigraph.

There is no doubt that Eliot refers in the epigraph, "A penny for the Old Guy", to Guy Fawkes, the Catholic conspirator executed for his part in a failed plot to blow up Parliament in 1605. Stuffed effigies of Fawkes were created by English children to beg for money prior to Bonfire Night, on 5th November, which commemorates the failed plot. This English tradition of children begging coins for their 'Guy' was recorded as early as 1826, with "the boys […] running up to passengers hat in hand, with 'pray remember Guy! please to remember Guy'".[49] The directness and therefore limitation of this reference has been perpetuated by scholars such as Grover Smith, who explains simply in his study of Eliot's sources and meanings that "'A penny for the Old Guy' is the formula by which children solicit money for fireworks on Guy Fawkes Day, the fifth of November".[50] Similarly Jane Worthington, in her study of Eliot's epigraphs, writes that "On Guy Fawkes Day children carrying straw effigies of Guy Fawkes about the streets cry out 'A penny for the Old Guy'. With the pennies thus collected, they buy firecrackers, carry their effigies to a large bonfire, and there 'with a bang' clear the world of Guy Fawkes for at least another year".[51]

However, bounding the reference in this way is flawed. Those who experienced this now-disappearing activity confirm that the children's request was "A penny for the guy?", not for "the *old* guy". Eliot would not have heard the cry as a child himself, as the tradition, connected as it is to English history, is not pursued in the United States. Iona and Peter Opie, who assiduously recorded the games and chants of English schoolchildren, cite four different rhymes relating to Guy Fawkes, and none relating to the begging include the word 'old'. They refer instead in their description to "the humble penny-for-the-guy city child".[52] The word 'old' was not used.

The insertion by Eliot of the capitalised word "Old" into his epigraph therefore supports a further connection, to Charon, the old ferryman of classical mythology who conveys the dead to the Underworld across the river Styx. One is particularly struck by the auditory echo of the Styx in Eliot's remark to his translator, that the latter holds "the right stick to beat the old guy with". Charon is indeed an 'Old Guy'; in the *Aeneid* Virgil emphasises his age, his gruesome appearance, squalid clothing and unkempt beard, reminiscent of the old man in 'Prufrock's Pervigilium' (See Chapter 2). Virgil does however remind us that, unlike a man, "a god's old age is fresh and strong".[53] Further into his own crossing passage cited above, Dante describes Charon as "*un vecchio bianco per antico pelo*"—"an old man, white, with ancient hair".[54] He is to lead his passengers into "eternal darkness",[55] a form of blindness, whereas Charon himself has "eyes of glowing coal",[56] a vivid contrast to the "sightless" hollow men.

The "tumid river" upon whose beach the hollow men are gathered is therefore the Styx. This was confirmed by Eliot in a letter: "I certainly had the Styx in mind and perhaps with a rather more antique than Dantesque association", suggesting the Virgilian source; although with customary obfuscation Eliot continued "but it is not intended to be a very precise reference".[57] This river also resonates with that which runs through Joseph Conrad's *The Heart of Darkness*, the work which provides the poem's other epigraph.

In order to convey his passengers across the Styx—"those who have crossed […] to death's other Kingdom"[58]—Charon demanded a coin in payment; the Romans would place one in a corpse's mouth. Hence the penny which is begged in the epigraph for the Old Guy. This recites

Eliot's familiar tension between youth and old age, for it is the begging efforts of children which will provide the hollow men's passage. Two definitively old men are presented as controlling figures over the lives of the hollow men—Charon and Charlatan. These orthographically neighbouring old men determine the life and death of the stuffed characters. The ability of those characters to act, to live, is determined by one old man, while their ability to die is determined by the other. Old age is granted its ultimate power over life and death.

In a lecture delivered at Cambridge in November 1924, coinciding with publication of the first section of what would become 'The Hollow Men', Eliot spoke of "the Dissolution of Christianity in protestant Europe, of the relaxing of the Christian system of the various needs of man", and said "This is I think the great distress of the modern world, that it is neither Christian nor definitely something else".[59] Lacking a redemptive response to the stasis of old age, 'Gerontion', *The Waste Land* and 'The Hollow Men' might be seen to occupy that void. Eliot's imminent conversion to the Anglican church would see his poetry incorporate that "Christian system" and its potential for salvation to be a conclusion to old age.

Notes

1. Eliot, T.S. 2015. *The Poems of T.S. Eliot Vols I and II*. Ed. Christopher Ricks and Jim McCue. Faber and Faber, 1, 74.
2. Petronius, *Satyricon*, sec. 48.
3. Murphy, Russell Elliott. 2007. *Critical Companion to T.S. Eliot: A Literary Reference to His Life and Work*. New York, Facts on File, 433.
4. Pound, Ezra. 1987. *The Cantos of Ezra Pound*. Fourth Collected Edition, Faber and Faber, LXIV.
5. Eliot, *Poems* 1, 84.
6. Eliot, *Poems* 1, 324.
7. Eliot, T.S., Letter to Emily Hale, 2nd March 1931.
8. Vallin, J. 1991. 'Mortality in Europe from 1720 to 1914: Long-Term Trends and Changes in Patterns by Age and Sex', in *The*

Decline of Mortality in Europe, ed. Schofield, R., Reher, D., and Bideau, A., Clarendon Press, 38–67 https://www.sociostudies.org/almanac/articles/the_demographic_transition/.

9. Eliot, *Poems* 1, 87.
10. Eliot, T.S. 1939. *The Family Reunion*, Faber & Faber, Part 1, Scene 1.
11. Patel, Dhaval, MD, Joel Steinberg, MD, and Pragnesh Patel, MD., 'Insomnia in the Elderly: A Review', *Journal of Clinical Sleep Medicine*,Volume 14, Issue 06, 1018 https://doi.org/10.5664/jcsm.7172.
12. ibid., 1018.
13. Athenaeus. *The Learned Banqueters*. Volume I: Books 1–3.106e.
14. Améry, Jean. 1994. *On Aging: Revolt and Resignation*. Trans. John D. Barlow, Indiana University Press, 55-6.
15. Kyd, Thomas. 1959. *The Spanish Tragedy*. Ed. Philip Edwards, Methuen. Act III, Sc XI.
16. Kyd, *The Spanish Tragedy,* Act IV, Sc I.
17. Eliot, *Poems* 1, 74.
18. Eliot, T.S. 1971. *The Waste Land: A Facsimile and Transcript of the Original Drafts*. Ed. Valerie Eliot, Faber and Faber, 47.
19. Sophocles, *Oedipus Tyrannus*, 460.
20. Ovid, *Metamorphoses*, III, 511-527.
21. Callimachus. *Hymn V, 'On the Bath of Pallas'*, 49.
22. Ibid., 82.
23. Ibid., 101.
24. Hesiod, *Melampodia*, Fragment 2.
25. Callimachus, 128.
26. Sophocles, *Antigone*, 991, 1031, 1045.
27. Ibid., 1084.
28. Smith, Grover. 1956. *T.S. Eliot's Poetry and Plays: A Study in Sources and Meaning*. University of Chicago, 88.
29. Schmidt, A.V.C. 1976. 'Eliot, Swinburne and Dante: A Note on 'The Waste Land' Lines 215-248', *Notes and Queries*, Volume 23, Issue 1, January 1976, 17–18, https://doi.org/10.1093/nq/23-1-17.
30. Swinburne, Algernon Charles. 1880. *Songs Before Sunrise*. Chatto & Windus, 207.
31. Eliot, T.S. 2022. *The Complete Prose of T.S. Eliot: The Critical Edition Vols 1-8*. Ed. Schuchard, Ronald, Brooker, Jewel Spears,

Cuda, Anthony, Dickey, Frances, Formichelli, Jennifer, Harding, Jason, Javadi, Iman, Stayer, Jayme, and Chinitz, David. The Johns Hopkins University Press and Faber & Faber Ltd. http://muse.jhu.edu/resource/1. 1, 478.
32. Eliot, *Prose* 6, 262.
33. Tennyson, Lord Alfred. 1885, *Tiresias and Other Poems*. Macmillan, 5–18.
34. Habib, M A R, 1998, 'Writing of The Suffering of Tiresias in Greek Tragedy', *The Bucknell Review*, Vol. 42, Iss. 2, 22.
35. Eliot, *Poems* 1, 336.
36. Oxford English Dictionary, "dug (n.1), sense a," June 2025, https://doi.org/10.1093/OED/6936291848.
37. Eliot, *Poems* 1, 74.
38. Richardson, John, 'Age and Youth in 'The Waste Land', *English: Journal of the English Association*, Volume 39, Issue 165, Autumn 1990,, 215.
39. Richards, I.A. 1949. *Principles of Literary Criticism*. Routledge & Kegan Paul, 295.
40. Eliot, T. S. 2022. *The Complete Prose of T. S. Eliot: The Critical Edition Vols 1–8*. Ed. Schuchard, Ronald, Brooker, Jewel Spears, Cuda, Anthony, Dickey, Frances, Formichelli, Jennifer, Harding, Jason, Javadi, Iman, Stayer, Jayme and Chinitz, David. The Johns Hopkins University Press and Faber & Faber Ltd, http://muse.jhu.edu/resource/14, 226.
41. Richardson, 225.
42. Eliot, *Poems* 1, 84.
43. Eliot, *Poems* I, 836.
44. Eliot, *Poems* 1, 139.
45. Dante, *Inferno*, Canto III, 47.
46. Symes, Gordon. 1951. 'T.S. Eliot and Old Age', *The Fortnightly Review*, Vol CLXIX Jan-June 1951, 191.
47. Eliot, *Poems* I, 717.
48. Ibid.
49. Hone, William, 1826, *The Every-day Book and Table Book*, William Tegg & Co, I, 1431.
50. Smith, 103.
51. Worthington, Jane, 'The Epigraphs to the Poetry of T. S. Eliot', *American Literature*, Mar., 1949, Vol. 21, No. 1, 15.

52. Opie, Iona and Peter. 1959. *The Lore and Language of Schoolchildren.* Oxford University Press, 281.
53. Virgil, *Aeneid*, Book 6, 297.
54. Dante, *Inferno*, Canto III, 83.
55. Ibid., 87.
56. Ibid., 109.
57. Eliot, T.S. 2009–2025 *The Letters of T.S. Eliot Vols 1–10* Ed. Valerie Eliot, Hugh Haughton and John Haffenden. Faber & Faber 10, 350.
58. Eliot, *Poems* 1, 81.
59. Eliot, *Prose* 2, 555.

CHAPTER 6

"The Agèd Eagle": 1927–1935

Abstract Following his conversion, Eliot retains elderly personae in his poetry, in order to address the new perspectives which his faith presents, and contrast those with the perspectives of the old men he had depicted previously, and with the philosophy of lifelong service in which he had been raised. Both Simeon and the Magus are old men at the end of both their eras and their lives, engendering sympathy for their wish to depart. In 'Ash-Wednesday' Eliot expresses bitterness, frustration and resignation as he attributes failures of his own to senescence, and he revisits the image of old age which he has presented in the past in order to leave it behind. He sees himself now as an "agèd eagle", in a position of seniority from which he can castigate the behaviour of another ageing literary figure in 'Lines for an Old Man'—although other such literary figures question Eliot's assumption of senior status at the age of forty-two.

Keywords Religious conversion · Journey of the Magi · Song for Simeon · Ash Wednesday · Sweeney Agonistes · Lines for an Old Man

In June 1927, Eliot was baptised and confirmed into the Church of England, and the Ariel poems which Eliot wrote immediately following his conversion can be seen to reflect aspects of his own spiritual journey from agnosticism to affirmation. It is therefore significant that, in his

P. Keers, *T.S. Eliot*, Modern and Contemporary Poetry and Poetics,
https://doi.org/10.1007/978-3-032-25050-6_6

first two Ariel poems, 'Journey of the Magi' (1927) and 'A Song for Simeon' (1928), the narrators are old men.[1] He wrote to his mother after his conversion confirming his personal conviction that he would see her in another life,[2] and this spiritual belief would be reflected through elderly characters approaching natural death. By retaining elderly personae he could address the new perspectives which his conversion presented, and contrast those with the perspectives of the old men he had depicted previously.

At the heart of both 'Journey of the Magi' and 'A Song for Simeon' is the Birth of Christ, the event which to Eliot was the most significant in the Gospels; in his introduction to *Revelation*, a collection of essays by theologians, he would subsequently declare that all Christian revelation is to be understood in relation to the Incarnation.[3] In 'Journey of the Magi', the age of the narrating Magus is revealed when he states that the events he remembers took place a long time ago, the poem being framed as the recollections of an old man. This is not just a repeat of the poetic device Eliot first employed in his graduation poems; as Gerontion's old age embodied the history of Western culture, so the elderly Magus's consciousness can contain the history of Christian culture. That includes the words of Lancelot Andrewes, the priest whose words open the poem, but whose sermons take place long after the Magus's death, which we therefore understand in a manner that the Magus cannot.

The Biblical description of the Magi refers to them simply as "wise men from the east"[4]; there is no indication of their age. Again Eliot has imposed old age, and its consequent perspective, upon a character whose age has not been specified in the original source material. In an echo of the traditional alternative description of the Magi as 'kings', Eliot describes their privileged home life, with its palaces and concubines, not only contrasting with the rigours of their journey, but establishing a social status counter to his previous depictions of old men in gutters and decayed houses.

Eliot has also appended the adjective 'old' to a white horse which the Magus sees. There are two Biblical references to a 'white horse', both in the Book of Revelation,[5] potential portents of a Second Coming, but neither of the horses are described as 'old'. However, as Eliot's old white horse gallops away, the term resonates with the dissociation of the 'old dispensation' from the Magus.

However much time has passed since his journey, the Magus dictates his experience through recollection in old age, and his status accords him the privilege of setting down his words for posterity, unlike Gerontion whose musings remain impermanent tenant thoughts. Although we hear of the unease which the Magus suffers upon returning to his kingdom, there is no attempt in his record to share or explore his experience in the subsequent years with the people he now regards as alien; his recollections are solipsistic. The clarity and forcefulness of the Magus's repeated instruction to the scribe to set down his dictation counters any suggestion that the unease to which the old man confesses may be a symptom of senility. Instead, the absence of the kinds of geriatric deterioration previously depicted by Eliot sanctions our more common association of the Magi's old age with wisdom.

Yet the reader possesses a wisdom which the agèd Magus still lacks. We comprehend the significance of the Christian symbols, the three trees upon a hill, or the pieces of silver, which the wise old man observed but does not understand. Despite the status and significance of the Magus, Eliot ostensibly places us, as well as himself, as he has done previously, in a superior position to an agèd man. But that also positions us among those at whom Gerontion scoffs, who would see signs and take them for wonders. The Magus correctly interprets the most important sign—the Birth—despite the agony it causes him. He is not suffering the confusion of senility, but unease at the consequences of his understanding, and the wonder which he has experienced is more significant than any we feel at the presence of the unacknowledged scriptural signs. Our respect for the elderly Magus is not based upon his regal status, nor the arduous journey he has undertaken, but upon the revelation he has experienced. At the end of his life, we do acknowledge this old man's superiority, over those remaining in the previous dispensation, and over ourselves.

The salvation of both the Magus and Simeon is rooted in their *sight* of the Incarnation. Eliot's previous association of old age and blindness, both literal and metaphorical, is discarded after his conversion, to be replaced with eyes which, like Eliot's own, "have seen thy salvation".[6] Sight has replaced blindness in Eliot's old men; the Magus has seen birth and death, Simeon sees the Word which cannot yet speak. The eyes of the old men have now been opened to, as Eliot described it, "the supernatural end – beatitude – for those who have the eyes to see it".[7] This revelation inevitably causes dissatisfaction with their existing systems, the dispensations which have existed throughout long and, from the descriptions of the Magus's summer palaces and Simeon's honour, satisfying lives. It is conventional for old men to bemoan the manner in which society has changed during their lifetime; having seen beyond such temporal concerns, these old men are exercised about leaving that dispensation behind.

The Magus's concluding statement, that "I should be glad of another death", resonates on many levels, incorporating the Christian theology to come, linking birth and rebirth, Incarnation and Crucifixion, and the desire for a Second Coming, as well as the Magus's own end. Projecting into the Magus's old age enables Eliot to explore the profundity of death as a prospect imminent in the narrator's present; it is the proximity of death which compels the Magus to demand that the scribe record his realisation of the unity of Death and Birth. This awareness, together with the dissolution of a lifetime's dispensation, and the prospect of relief in death, are themes Eliot pursues further in 'A Song for Simeon'.

Both old men are presented at the end of their era and the end of their life. Parallels are drawn between the ageing of old dispensations, and the ageing of old men—both will now pass. The Magus cannot foresee the future, but acknowledges the passing of his past, while Simeon wishes to be part of the future, but knows that he cannot. Tired of their old lives, both wish to depart from their present.

In 'A Song for Simeon', Simeon's old age is openly stated—he is eighty, the first of Eliot's old men whose age is declared. Eliot's chronology of old age has changed significantly since *The Waste Land*, in which Marie was old at forty-two. In 'The Cultivation of Christmas Trees' Eliot will write of the eightieth Christmas being the last; Simeon hopes that this will be the case for him. The poem is set in the season of winter, and in the winter of Simeon's life.

The character of Simeon is taken from the Gospel of Luke, which relates how Simeon is present as Mary and Joseph present the infant Jesus in the temple.[8] It had been revealed to Simeon by the Holy Ghost that "he should not see death before he had seen the Lord's Christ". Upon seeing the child, Simeon takes him in his arms, and prays:

> Lord, now lettest thou thy servant depart in peace, according to thy word:
> For mine eyes have seen thy salvation[9]

These lines became the prayer known as the *Nunc dimittis*, part of the office for Compline in the Roman Catholic tradition, and Evening Prayer in the Anglican church. They are sometimes called the Canticle, or Song, of Simeon. Eliot's title replaces 'of' with 'for', opening up an ambiguity as to whether his poem is written to be expressed *by* Simeon, or is a lament *for* him. Answering a later query about a similar ambiguity in the title of 'Lines for an Old Man', Eliot said that in the latter case he meant 'lines in the mouth of'.[10] Here, the ambiguity means that we approach the 'song' both hearkening to, and lamenting for, Simeon's situation.

Eliot's grandfather, William Greenleaf Eliot, who exerted significant influence over Eliot's childhood, wrote a poem titled 'Nunc Dimittis' in celebration of his seventy-sixth birthday. It was printed in his biography, written by Eliot's mother Charlotte.[11] At the time of writing 'A Song for Simeon', Eliot was conscious of his mother's extremely frail condition, and less than a year prior to her death in September 1929 he had pre-empted in a letter that she was unlikely to survive another winter,[12] indicating the presence of her end of life in his thoughts. Eliot, notoriously resistant to biographical interpretations of his work, would later deny that he was thinking of his grandfather when he wrote 'A Song for Simeon'.[13] However, we have seen how Eliot readily acknowledged his grandfather's influence over his upbringing (Chapter 2), with his values and his 'Nunc Dimittis' memorialised by Eliot's ailing mother; his grandfather was among family photographs on Eliot's sitting room wall in this period[14]; and significantly, Eliot's Simeon refers to "my children's children", rendering him a grandfather, although this is not evidenced in the Gospel.

The opening stanza of William Greenleaf Eliot's poem describes him suffering from "failing strength", and wondering when he might utter the *Nunc dimittis* lines of Luke: "Now lettest thou thy servant, Lord, Depart in peace", to claim the "kind award" of his place inside the gates of heaven. However, his response in the subsequent stanza is emphatic. He quashes his own plea for death—"Be still, my heart!"—arguing that God's silence communicates that one's place is here on earth, where one should continue "doing or suffering" in old age until death comes.[15] This is the Puritan morality which Eliot remembered from his childhood, the philosophy of service until one's very end outlined in Chapter 2, and summarised in his grandfather's poem as "If daylight fades, work while the twilight lasts".[16]

Eliot was aware that a legacy of mission dominated his grandfather's faith: "Although his piety was great, his interests and activities lay rather in the direction of philanthropy and public service, and not in that of abstract thought or the contemplative life".[17] The latter, however, was the course which Eliot himself admired and pursued. Even before his conversion, in Arthur Symons' *The Symbolist Movement in Literature* (see Chapter 3) he had encountered Symons' idea that while a man might "in his lower moments"[18] serve God in action, one's ultimate attainment was "an incommunicable ecstasy", whereupon "we realise the infinite insignificance of action".[19]

Thus, in contradiction to William Greenleaf Eliot's edict, Simeon echoes the Sibyl in requesting release from old age in death. His self-assertion counters the self-denial of Eliot's grandfather. Simeon's religious, philanthropic and administrative "doings" are done and, unlike Eliot's grandfather, who believes to the end that his "task is not yet finished", Simeon lays claim to his departure. He has "eighty years and no to-morrow", the length of the former term contrasting with the abruptness of the latter in a phrase which juxtaposes time past and time future, suggesting immediate temporal termination, and therefore also a release from age, time's constant measurement of life. If Simeon has no tomorrow he cannot work on in the manner Eliot's grandfather espouses; like the dust and feathers imaged in the poem, Simeon's physical existence is no longer a substantial presence. Like Gerontion, he awaits the final wind.

As with other Eliot characters, there is no source evidence that Simeon was an old man. Luke states that Simeon had been "waiting for the consolation of Israel", but not for how long. It was prophesied that he would not die *before* seeing the Christ, so it would presumably be possible for him to do so in middle age, and then subsequently live on after the experience for many further years. Six verses later, Luke refers to a prophetess, Anna, as being "of a great age",[20] and so is clearly aware when the age of a subject is significant and should be stated. Simeon's old age has been specified only by Eliot.

It means that Simeon can look back upon a life, and one of material success, wealthy enough to give alms to those who come to his door, and with the status to grant honours as well as to receive them. This is not the decrepit old man sitting in the gutter, or the tenant in a decayed house. Simeon has done philanthropic work, and followed the Laws; Luke describes him as "just and devout". Simeon defines himself positively by what he has done, whereas Gerontion defines himself by what he has *not* done and Phlebas forgets his mercantile achievements. In Joseph Conrad's *Heart of Darkness*, immediately before the lines which Eliot chose but finally rejected as the epigraph to *The Waste Land*, the narrator Marlow comes across a dozing Kurtz, who just before his moment of supreme knowledge mutters what could be a summary of Simeon's circumstances: "Live rightly, die, die…".[21]

Depicting Simeon as elderly enables Eliot to instil additional theological significance into his passing by drawing upon the writings of St Paul, who urges "That ye put off… the old man, which is corrupt according to the deceitful lusts; / And be renewed in the spirit of your mind".[22] For St Paul, the term "old man" symbolises a life apart from God. The renewal which can be achieved when this symbolic "old man" is "put off" can be reached through death; "knowing this, that our old man is crucified with Him, that the body of sin might be destroyed […] For he that is dead is freed from sin".[23]

William Greenleaf Eliot was seventy-six when he wrote his 'Nunc Dimittis'; Simeon is made a directly comparable eighty. This means that Simeon's contemplative perspective, which Eliot shares, can be directly contrasted to the service ethos of Eliot's grandfather, without any potential complications arising from either a difference in years or distance from death clouding the comparison. In their responses to the *Nunc dimittis*, the two grandfathers can reflect the transition which Eliot himself has made, from the Unitarian religious values of service in which he was raised to the philosophical values of the Anglo-Catholicism to which he has now converted. Once a foil to his youthful *carpe diem* ideas, his grandfather's principles continue to prove productive to Eliot's thinking. In a footnote to a *New Criterion* article in January 1926, Eliot modified an aphorism by Guillaume Apollinaire, and wrote "*on porte partout le cadavre de son grand-père*"—You carry everywhere the corpse of your grandfather.[24]

Unlike that old man Tiresias, Simeon has not foresuffered all—far from it—but he similarly foretells the suffering which will occur after him. Rather than solely looking back in recollection, this old man looks forward. Nor is his vision solipsistic, for while he accepts the experiences absent from his own future—martyrdom, ecstasy and the ultimate vision—he is anxious about the experiences of others to come. While Tiresias remains detached from and contemptuous of the activities of the young, Simeon displays concern for his grandchildren. And while he addresses his spiritual future, he also reflects the physical concerns of old age. As Symes observes, "For while old age, by its release from sensuous distraction, might seem to be particularly receptive to the counsels of the spirit, it is also correspondingly subject to a profound longing for sleep and conclusion".[25] With his admirable efforts to 'live rightly', and his unselfish framing of a just terminal request, we at last encounter an old man at the end of his life who fully engenders our sympathy.

'Ash-Wednesday' (1930) is Eliot's first poem presented *as* an old man, following years of poetic presentations *of* old men. Eliot's previous monologues either named the narrators or placed them in circumstances which distanced them from their creator. However, as the first-person narrator of 'Ash-Wednesday' was identified spiritually with the post-conversion Eliot at the time of publication, the confessional nature of the poem also alerted his audience that Eliot at forty-one might be declaring a concern about his own ageing. When, at half that age, he wrote of Prufrock's anxiety at growing old, none would have imagined that to be Eliot's personal concern; hence Wyndham Lewis's confusion upon meeting the poem's young author (see Chapter 2). Now, that concern with age could be recognised; as it was, notably and negatively, by the critic Edmund Wilson, who wrote: "I am made a little tired at hearing Eliot, only in his early forties, present himself as an 'agèd eagle' who asks why he should make the effort to stretch his wings".[26]

Letters from this period confirm Eliot's concerns about ageing. In 1927 he wrote to his mother: "I have still – though I know that I am getting old – ambitions of things to be done".[27] He was thirty-eight. In November 1930, he confesses to Emily Hale that, in the period following *The Waste Land*, his marital love for Vivienne had "failed utterly"; he had considered the justification for a separation, but "It is so hard to persuade oneself that one's work is important enough to sacrifice another person to it – at least, it is hard when one is middle-aged". And he writes that, in his late thirties, "I resigned myself to celibate old age".[28]

Acknowledging Eliot's concern with age provides a particularly personal approach to 'Ash-Wednesday', and the poem's powerful images of senescence. The opening line of 'Ash-Wednesday', "Because I do not hope to turn again", is taken directly from the Ballatetta written by Dante's 'first friend' Guido Cavalcanti, 'Perch'io non spero'. Eliot used that title when this section of 'Ash-Wednesday' was first published independently in Spring 1928. Gordon Symes asserts that Cavalcanti's poem "...was written in exile, and on the verge of death, and may therefore be taken to signify that acute sense of deprivation with which old age is first apprehended".[29]

It is not possible to confirm just how closely it was written to Cavalcanti's death in 1301, after his return to Tuscany in sickness from exile, but the Ballatetta expresses a resigned proximity to death which echoes that of Gerontion and Simeon:

Tu senti, ballatetta, che la morte
mi stringe sì, che vita m'abbandona

Thou knowest, Ballatetta, that death
Lays his hand upon me, whom life has forsaken

It is in this state of mind that Eliot considers whether he can hope to "turn". The theological sense of 'turning' will become a significant theme of the poem; there are repeated Biblical references to 'turning' to God,[30] and a line from Joel 2.12, "Turn ye even to me, saith the Lord, with all your heart", begins the epistle for 'The First Day of Lent Commonly Called Ash Wednesday' in *The Book of Common Prayer*.[31]

However, when the narrator perceives himself as agèd, there arises an issue of whether he considers that he has sufficient life remaining to him in which to 'turn', and whether old age is a hindrance to turning, either because of the weight of experience, the erosion of ability or the lack of time. The narrator has abandoned the ambition to succeed. Shakespeare identified "this man's art and that man's scope" as enviable attributes in his Sonnet 29. Eliot negates the verb 'to strive' in order to reject further action in their pursuit. At this point in life, the narrator's power and status ("reign") has "vanished", indicating that it had been present in the past; identifying the narrator with Eliot suggests this is creative power, but by replacing Shakespeare's 'art' with 'gift' he avoids voicing envy of others' creative talents, and instead generalises their desirable attributes. Combining bitterness, frustration and resignation as he attributes his failure to old age, Eliot wonders:

(Why should the agèd eagle stretch its wings?) [32]

Eliot would later suggest that this description was "the kind of exaggeration that goes with immaturity, seeing oneself as older than one is".[33] There is a truculence in the line, typical of irritable old men. But together with his assumed age, the persona reflects Eliot's status and reputation, his power and reign at the time. Eagles combine with trumpets in both 'A Cooking Egg' and 'Coriolan', where they provide symbols of status and achievement.

The symbol also reflects fears about a geriatric waning of his creative status. In a 1932 lecture, Eliot would question the less successful works of Wordsworth's old age:

> Whether his power and inspiration remained with him to the end is, alas, not even doubtful. But Wordsworth had no ghastly shadows at his back, no Eumenides to pursue him; or if he did, he gave no sign and took no notice; and he went droning on the still sad music of infirmity to the verge of the grave. His inspiration never having been of that sudden, fitful and terrifying kind that visited Coleridge, he was never, apparently, troubled by the consciousness of having lost it. As André Gide's Prometheus said, in the lecture which he gave before a large audience in Paris: *Il faut avoir un aigle*. Coleridge remained in contact with his eagle.[34]

Prometheus's lecture takes place in Gide's novel, *Le Prométhée mal enchaîné*, in which an eagle replaces the vulture of classical mythology, and regularly eats the liver of a contemporary Prometheus. To Gide's Prometheus, the eagle represents not just inspiration, but "*Notre aigle est notre raison d'être, Messieurs*".[35] The wings of Eliot's eagle should impel him forward after stretching, but these wings are no longer capable of flight. They are now merely "vans", an obsolete and "chiefly poetic"[36] term for wings, once used by Milton, and by Dryden in his translation of Ovid's *Metamorphoses*, where a "new-dissembled Eagle… stretch'd his Vans in vain / His Vans no longer cou'd his Flight sustain".[37] The term therefore has a resonance with past poetic achievements, and with Ovid, whose Tiresias legend Eliot drew upon for *The Waste Land*; but the term's very obsolescence reinforces the agèd nature of this eagle. These vans merely "beat the air", hence the pointlessness of stretching, preparatory to a flight which he fears cannot take place, his "vanished" creative power, the reign over words to which he has been accustomed, anticipating homophonically the subsequent "vans".

The potential for rebirth is, however, also contained within the agèd eagle. On his proof copy of the poem, Eliot's colleague John Hayward noted both Dante's *Purgatorio*, and medieval Bestiaries.[38] The thirteenth-century Guillame le Clerc recorded in his *Bestiaire Divin*:

> The eagle is the king of birds. / When it is old it is renewed again / Through its most wonderful nature. / A fountain clear and pure, / Where the water springs fresh and bubbling, / When the sun is shining bright / The eagle seeks, when it is old / And its eyes are very dim / And

each wing heavy and feeble. / In the air above this fountain / It begins to mount very high / In the face of the sun shining bright. / When up there it meets the heat, / Its eyes it fastens on the glare / Of the sun and gazes so hard on it / That it reckons to be all on fire. / There in the heat it burns / Its eyes and its wings likewise; / Then it flies down into the fountain, / There where the water is most fresh and pure, / And dips and bathes three times, / Until it is – be well assured – / All fresh and all renewed / And from its old age healed.[39]

This legend associating agèd eagles with rebirth underlies the dream which Dante records at the beginning of *Purgatorio* Canto IX (seen as providing the "three stair" structure which Eliot uses in section III of 'Ash-Wednesday'). Dante dreams of an eagle which carries him up to the sun, where he and the eagle burn in "the visionary flame so scorched that needs was my slumber broken".[40]

The dream prophecies Dante's own purgatorial journey to come; he will journey up through Purgatory to Paradise through the refining fire—the *foco che gli affina* quoted at the end of *The Waste Land*. St John of the Cross draws upon the same legend, referring again to the Pauline concept of the "old man":

> In this way God makes it to die to all that is not naturally God, so that, once it is stripped and denuded of its former skin, He may begin to clothe it anew. And thus its youth is renewed like the eagle's and it is clothed with the New man, which, as the Apostle says, is created according to God.[41]

Eliot was also to draw upon this in *The Family Reunion*, where the character Mary asks "And what of the terrified spirit / Compelled to be reborn / To rise toward the violent sun".[42] This introduces compulsion into the rebirth cycle, with an even older Eliot acknowledging fear of a process involving scorching fire, rather than the tranquil departure from life commonly hoped for in old age. Symes believes that "T.S. Eliot is here speaking 'emblematically' – not personally – of that universal antipathy to regeneration which hardens with age, to the extent that death begins to be preferred to change".[43]

The agèd eagle brings together in a single symbol Eliot's deliberations over his spiritual journey, his concerns about old age, and his fears about his creative status in senescence. Perhaps because of its significance he attempted to be dismissive of it when questioned in his final years, saying that "I am afraid my mind was very empty of allusions when I used the phrase 'agèd eagle'. It just came that way".[44] But after his death Eliot's friend, the literary critic I.A. Richards, confirmed that "The agèd eagle persona was dear to him".[45]

Counter to the dignity of the agèd eagle persona, and more akin to his previous depictions of old men, the poem proceeds to present one of the most vivid images of old age in Eliot's oeuvre. Section III is structured around the climbing of three stairs, echoing both Dante's *Purgatorio*, and the "saint's stair" which Simeon foresees being mounted by subsequent generations. In 'Ash-Wednesday':

> ...the stair was dark,
> Damp, jaggèd, like an old man's mouth drivelling, beyond repair.[46]

The older sense of 'drivelling', to mean slavering or drooling, the dribbling of saliva from the mouth, gives the image its repellance. It echoes the "*baveux*" or drooling waiter in 'Dans le Restaurant'.[47] As with the use of 'vans', employing an outdated usage underscores the age of the object. Old age is being associated with another physical decline—oral; muscular control has been lost, leaving the mouth "damp", which has associations with stagnation and mould. 'Jaggèd' teeth suggests broken, and recalls the "Dead mountain mouth of carious teeth that cannot spit"[48] in *The Waste Land*. The contemporary use of the word 'drivelling', to speak or write in a foolish manner, provides a further sense of deterioration in later life, in this case of expression. Human intervention can do nothing to repair these consequences of ageing. This old man's mouth can offer nothing but drivel.

To provide an image of a corporeal world which needs to be discarded, Eliot has turned again to old age. While the Pauline concept of transcending a spiritual 'old man' is again echoed in the narrator's climbing above the geriatric image, the spiritual symbol is given vivid material presence. The dual senses of 'drivelling' combine to present a deterioration of both body and mind, which cannot be repaired in the physical realm, and which must be left behind, as the narrator climbs above it on the stair from purgatory. Eliot is not just rising above *an* image of deterioration—he is rising above *his* image of it which he has utilised for over fifteen years. This is not to abandon old age as a poetic device, or as a concern, as we shall see, but it is to rise above its depictions as pitiable, frightening or repellent. These will not recur in Eliot's subsequent images.

The old man is immediately followed by the image of another agèd creature, a shark. Much as Gerontion's consciousness spanned the entire history of Western culture, and the Magus's that of Christian culture, so the shark as a species spans the history of earthly life; shark fossils have been found from the Late Ordivician Period, 450 million years ago, while the oldest known group of definably modern sharks date from the Early Jurassic Period.[49] Eliot's historic sense has often led to the incorporation of references to (and from) recorded history; the shark goes beyond this, into a presence from *un*recorded history. In climbing above this creature, the narrator distances himself from historic life on earth.

In addition, individual sharks are extremely long-lived. An agèd shark may be more than 130 years old; yet its threat, unlike that of other agèd creatures, persists—it remains "toothed". (Unlike man, a shark's teeth are naturally replaced—or 'repaired'—thoughout the shark's lifetime whenever lost or broken.) The teeth and gullet of the agèd shark suggest consumption, as opposed to the ineffective broken teeth and failing mouth of the old man. Writing about 'Gerontion', Jewel Spears Brooker highlighted Eliot's image of Christ as a tiger, "an image of a ferocious, devouring presence as a metaphor for metaphysical absence [...] The absence of God is not mere absence; nihilism is an absence with claws and teeth, capable of inflicting immense harm".[50] This is equally applicable to the shark of 'Ash-Wednesday', a similarly devouring presence, with additional significances resonating from its old age.

In rising above these images, Eliot leaves behind his most negative representations of old age. While old age continues to be prevalent within his work, its representation and deployment in subsequent poetry is of a less negative and more contemplative nature.

The two fragments which formed *Sweeney Agonistes: Fragments of an Aristophanic Melodrama* were first published in the *Criterion* in October 1926 and January 1927, and only published together as a slim book in December 1932. The fragments themselves do not contribute to the present concern, but in March 1933, in response to a request to perform the work at Vassar College, Eliot provided new additional material, in which a character titled "Old Gentleman" enters the stage.[51]

The term "gentleman", coupled with Eliot's direction that he is in full evening dress, gives the character a class and status absent from his published descriptions of old 'men'. Apart from "old Tom" and the revellers excised from the drafted opening of *The Waste Land* (see Chapter 5), this is a new poetic portrayal for Eliot; Simeon's social status was stated, but not reflected in any physical description. Here the Old Gentleman "otherwise resembles closely Father Christmas", suggesting a long white beard, but also introducing an element of avuncularity, a characteristic of old age which will resurface significantly in the character of Old Possum. "In one hand he carries an empty champagne bottle, in the other an alarum [*sic*] clock". The old gentleman is preceded by "18 knocks like the angelus", a Catholic devotion commemorating the Incarnation of Christ. This recalls the couple in 'A Game of Chess' who are waiting for a knock upon their door to resolve their stasis.

"Good evening", he introduces himself, "My name is Time". Unaware of the direction referencing Father Christmas, the character would surely be perceived by an audience as Old Father Time, fresh from a New Year celebration with an empty champagne bottle and his clock. Old Father Time is often depicted, more akin to Eliot's earlier images of senescence, as bowed and decrepit, in contrast to the new-born baby of the New Year. His portrayal here in evening dress suggests a celebration at the passage of time, recalling the "undertakers ball" in 'Opera'.[52] He declares that his role is to "wait for the lost trains that bring in the last souls after midnight"; again there is a sense of transition taking place at a turning point of day and year—of time—being welcomed by an elderly figure.

The Old Gentleman asks whether Sweeney has any questions for him, the last of which is "When will the eagle get out of his barrel-roll?". Here Eliot's previously-employed eagle is pictured not in rising and rebirth, or even in agèd reluctance, but in what is commonly a terminal manoeuvre, offering no potential achievement but at risk of falling. Given Eliot's personal association with the eagle, the question could be read as an acknowledgement of a creative stasis. The Old Gentleman does not answer Sweeney's question in an elucidatory manner; he leaves, whereupon his alarum clock goes off.

This brief additional material indicates a transition in Eliot's presentation of old men. He does not exploit the opportunity to use decrepitude in the character of Father Time, but instead awards him the appearance and status of a 'gentleman'; having left behind his most negative imagery on the stair in 'Ash-Wednesday', we now see a more sympathetic portrayal of an old man, displaying status without physical or mental decline. Described overtly as "Old", the physical lifetime of the character, as with the pub ladies in *The Waste Land*, is used to represent the chronological passage of time. This device will become more significant as the two concepts, old age and time, interweave in *The Family Reunion* and *Four Quartets*.

In 1933, in a 'Critical Note' to *The Collected Poems of Harold Monro*, Eliot wrote that "The compensations for being a poet are grossly exaggerated; and they dwindle as one becomes older, and the shadows lengthen, and the solitude becomes harder to endure".[53] Aged forty-four, Eliot wrote from the United States to Harold Monro's widow Alida (who edited the volume), explaining that "America has not made any difference, except in making more real to me the fact [...] that I am getting old – I mean that my years of activity can be counted, and I have no more time now to waste".[54]

'Lines for an Old Man' (1935) appeared only in the *New English Weekly* before it was included as a 'Minor Poem' in Eliot's *Collected Poems: 1909–1935*. It has subsequently received little critical attention. Titles commencing "Lines..." were used on each of Eliot's little group of minor poems published as 'Five Finger Exercises' in the *Criterion* in January 1933. The title is appropriate to the relative slightness of those poems, suggesting easy, practice pieces for keeping one's hand in. Perhaps consequently, the draft title of this poem was 'Words for an Old Man', the use of 'words' differentiating it from those slighter 'lines'. After he had finally adopted 'Lines...', Eliot was asked "what was the exact meaning of the title? Was it 'Lines *in the mouth of* an old man?" and he replied, "Yes".[55] As with 'A Song for Simeon', the words are narrated by the poem's elderly subject, but without the sympathy engendered for Simeon. The title identifies the subject purely as an old man, elevating the significance of his age by defining him through old age alone.

The poem was dedicated in its draft version "to Stéphane Mallarmé", one of the French poets Eliot first encountered in *The Symbolist Movement in Literature* by Arthur Symons. The last line of the poem is a direct translation of a line from Mallarmé's poem 'M'introduire dans ton histoire' (1914): "*Dis si je ne suis pas joyeux*". (Drafts also included a translation of the subsequent line, "*Tonnerre et rubis aux moyeux*", which was later adapted for 'Burnt Norton' II, 1–2[56]).

Mallarmé was hugely influential on Eliot's pursuit of impersonality in poetry; Symons quotes Mallarmé's dictum that "The pure work [...] implies the elocutionary disappearance of the poet, who yields place to the words".[57] Eliot translates a line from Mallarmé's *Le Tombeau d'Edgar Poe*, "*donner un sens plus pur aux mots de la tribu*" into the impulse "To purify the dialect of the tribe" quoted in 'Little Gidding'[58], establishing Mallarmé's presence in the poem's 'compound ghost' of past masters.

Mallarmé held weekly *salons*, his "Tuesdays of the Rue de Rome" as Symons described them, "in which he gave himself freely to more than one generation"[59] Combined with the original dedication, these *salons* might be seen as the location for Eliot's old man's encounters with literary youth. However, Mallarmé should not be identified himself with the old man; Symons highlighted "the singular gentleness with which Mallarmé's intelligence moved", and in his description of the *salons* Mallarmé exhibited nothing of the caustic wit of Eliot's old man:

> One of the best talkers of our time, he was, unlike most other fine talkers, harmonious with his own theories in giving no monologues, in allowing every liberty to his guests.[60]

Victories of wit are all that remain to Eliot's old man. The self-satisfied triumph of old age over youth in 'Lines for an Old Man' is one of superficiality. The narrator has a self-image as a dominant creature, but is angry at the position to which life has led him, and from which, like the tiger trapped in a tiger-pit, he cannot break free. Nevertheless, when he scents an opportunity to conquer an opponent with his years of acquired wit, he poises for attack. It is a love-hate relationship with the young, whose defeats he needs to achieve in order to retain his sense of status. His "golden eye" emphasises his age by recalling the "agèd eagle" of 'Ash-Wednesday', and literally reflects the triumph of his cutting remarks over ignorant youth.

Eliot was suspicious of those elder poets who claimed to encourage the work of younger men. We have seen (Chapter 4) how Eliot did help and encourage a young poet himself, but when John Masefield, then Poet Laureate, declared in 1935 that older men wished only to encourage, advance and applaud young poets, Eliot challenged this in the *New English Weekly*:

> I am now old enough to recognise a qualified truth in what Mr. Masefield said: that is, I am sure that all elder poets *believe* that they wish to encourage and help younger ones, they *believe* that they really care for poetry, and not merely for what they write themselves. And they can really crave the pleasure of being able to enjoy and applaud work of younger

men. Only, their capacity for appreciation of anything new is apt to be very limited.[61]

That limited appreciation did not, however, justify immoderate criticism. Eliot's 1934 collection of lectures published as *After Strange Gods* contained attacks on contemporary writers including Lawrence, Hardy and Yeats. He insisted that the book went out of print, later saying that "there was a violent and intemperate tone about these lectures", and criticising himself for, in particular, "an intemperate attack on the people with whom I disagree".[62] The Old Man is unaware of the shame which Eliot will acknowledge as an old man himself in 'Little Gidding' as "things ill done and done to others' harm".[63] The Old Man may be representative of a type, or like the subjects of Eliot's other "Lines…" be based upon a personal experience; but while acknowledging his elderly literary character's fear of usurpation, Eliot would not condone his behaviour.

The concluding exclamation mark alerts us to the self-convincing nature of the Old Man's assertion of happiness, in the face of his corrosive irritability, bitterness and contempt. He believes that his superficial victories in old age compensate for his failings in life. Whereas youth behaving in the same manner might later repudiate such actions, the Old Man has no time in which to 'turn'; he is blind to the futility of these shallow temporal victories achieved by words alone, and falls back upon a complacent pleasure in the perceived personal triumphs which his words provide.

This old man should be seen in the context of Eliot's subsequent promulgation of the virtue of humility. In later works, Eliot will present humility as the greatest virtue—"The only wisdom we can hope to acquire / Is the wisdom of humility: humility is endless".[64]—and in his very last essay, published posthumously as 'To Criticize the Critic', he concludes by hoping that "as the critic grows older, his critical writings may be less fired by enthusiasm, but informed […] one hopes, by greater wisdom and humility".[65] His Old Man displays no such humility, and no intention of abandoning his pride; hence he has no prospect of redemption.

Old age had proved fruitful as a vehicle for spiritual reflections raised by Eliot's newly confirmed faith. The contrast between the portrayals of old men in this period and those in his earlier poetry expresses the transformation fostered by his conversion, and a spiritual consideration of old age which is becoming increasingly personal.

Notes

1. Another Ariel poem, 'The Cultivation of Christmas Trees', was also narrated by an old man, but was not published until 1954 and so is considered later.
2. Eliot, T.S. 2009–2025. *The Letters of T.S. Eliot Vols 1–10* Ed. Valerie Eliot, Hugh Haughton and John Haffenden. Faber & Faber 3, 647.
3. Eliot, T. S. 2022. *The Complete Prose of T. S. Eliot: The Critical Edition Vols 1-8*. Ed. Schuchard, Ronald, Brooker, Jewel Spears, Cuda, Anthony, Dickey, Frances, Formichelli, Jennifer, Harding, Jason, Javadi, Iman, Stayer, Jayme and Chinitz, David. The Johns Hopkins University Press and Faber & Faber Ltd, http://muse.jhu.edu/resource/1, 5, 472.
4. Matthew 2, i.
5. Revelation 6:2 and 19:11.
6. Luke 2:30.
7. Eliot *Prose* 5, 700.
8. Luke 2:25–35.
9. Luke 2:29–30.
10. *Eliot, T.S. 2015. The Poems of T.S. Eliot Vols I and II 1. Ed. Christopher Ricks and Jim McCue. Faber and Faber,* 1, 855.
11. Eliot, Charlotte C. 1904. *William Greenleaf Eliot: Minister, Educator, Philanthropist*, Houghton, Mifflin, 351.
12. Eliot, *Letters* 4, 125.
13. Eliot, *Poems* 1, 766.
14. Eliot, *Letters* 3, 647.
15. Eliot, Charlotte C., *William Greenleaf Eliot*, 351.
16. ibid.
17. Eliot, T.S., Letter to Emily Hale, 14th June 1932.
18. Symons, Arthur. 1899. *The Symbolist Movement in Literature*. William Heinemann, 173.
19. Symons, 175.
20. Luke 2:36.
21. Conrad, Joseph. 2006, *Heart of Darkness*. Norton Critical Edition, W.W. Norton & Co. Fourth Edition, 68.
22. Ephesians 4, 22–23.
23. Romans 6: 6–7.
24. Eliot, *Prose* 2, 765 n3.

25. Symes, Gordon. 1951. 'T.S. Eliot and Old Age', *The Fortnightly Review*, Vol CLXIX Jan-June 1951, 192.
26. Wilson, Edmund. 1931. *Axel's Castle*. Charles Scribner's Sons, 130.
27. Eliot, *Letters* 3, 648.
28. Eliot, Letters to Emily Hale, 3rd November 1930 and 1st May 1931.
29. Symes, 191.
30. Eliot, *Poems* 1, 734.
31. https://www.churchofengland.org/prayer-and-worship/worship-texts-and-resources/book-common-prayer/collects-epistles-and-gospels/ash.
32. Eliot, *Poems* 1, 87.
33. Gardner, Helen. 1958. 'The Agèd Eagle Spreads His Wings', Sunday Times, 21st September, issue 7062, 8.
34. Eliot, *Prose* 4, 626/7.
35. Gide, André. 1899. *Le Prométhée mal enchaíné*. Paris Société du Mercure de France, 93.
36. Van, N. (1)." Oxford English Dictionary, Oxford UP, March 2024 – 3. A wing; = fan n.1 4. Chiefly poetic. https://doi.org/10.1093/OED/1945564554
37. Ovid. *Metamorphoses*. trans. John Dryden, Book XII, 167.
38. Eliot, *Poems* 1, 736.
39. Le Clerc, Guillame, 'L'aigle', *Bestiaire divin*, 1210, Book XII.
40. Dante. 1901. *Purgatorio*. Canto IX, Temple Classics Edition, JM Dent & Sons, 33.
41. St John of the Cross. *The Dark Night of the Soul*. Chapter XIII, 11.
42. Eliot, T.S. 1939. *The Family Reunion*. Faber & Faber, Part I, Scene II.
43. Symes, 188.
44. Eliot, *Poems* 1, 737.
45. Richards, I.A., 'On TSE: Notes for a talk at the Institute of Contemporary Arts, London, June 29, 1965', in *The Sewanee Review*, Winter, 1966, 27.
46. Eliot, *Poems* 1, 91.
47. Eliot, *Poems* 1, 45.
48. Eliot, *Poems* 1, 68.

49. Davis, Josh, 'Shark evolution: a 450 million year timeline' Natural History Museum, London, https://www.nhm.ac.uk/discover/shark-evolution-a-450-million-year-timeline.html.
50. Brooker, Jewel Spears. 2006. 'Youth And Age In T. S. Eliot's Spiritual Development.' *Sewanee Theological Review 49.4* (2006): 473.
51. Eliot, *Poems* 1, 801.
52. Eliot, *Poems* 1, 236.
53. Eliot, *Prose* 4, 524.
54. Eliot, *Letters* 6, 552/3.
55. Tomlin, E.W.F. *T.S. Eliot: A Friendship*. Routledge, 1988, 66.
56. Eliot, *Poems* 2, 468.
57. Symons, 199.
58. Eliot, *Poems* 1, 205.
59. Ibid., 184.
60. Ibid., 184.
61. Eliot, *Prose* 5, 268.
62. Eliot, *Letters* 7, 245 n1.
63. Eliot, *Poems* 1, 205.
64. Eliot, *Poems* 1, 188.
65. Eliot, *Prose* 8, 469.

CHAPTER 7

"The Wisdom of Old Men": 1936–1946

Abstract In characters, images and, finally, personal confession, old age suffuses this period of significant dramatic, critical and poetic works. *The Family Reunion* contains Eliot's most realised elderly female characters, their age a personification of the passage of time, as the play considers the futility of attempting to control that passage. Old Possum presents elderly characters who, having lived full lives, now evince contentment and pride in their dotage. Finally in *Four Quartets*, Eliot identifies, in a manner sometimes critical, sometimes self-accusatory, a litany of failings and 'wrongs' in old men.

Keywords Family Reunion · Old Possum · Uncle Remus · Cats · East Coker · Little Gidding · Four Quartets

Having consistently projected a maturity greater than his years, it is perhaps not surprising that little change was noted in Eliot as he entered his fifties. "As in his thirties and forties he had looked mature", noted Robert Sencourt, "so now in his fifties he looked no older".[1] This stasis in appearance mirrors a similar consistency in character which Eliot had asserted: "Because as we grow old we all like to think that we have changed, developed and improved; people shrink from acknowledging that they are exactly the same at fifty as they were at twenty-five".[2]

P. Keers, *T.S. Eliot*, Modern and Contemporary Poetry and Poetics, https://doi.org/10.1007/978-3-032-25050-6_7

Eliot's elderly poetic personae wrestle in themselves with conflicting perspectives. In a dramatic format, these alternatives can be personified in separate characters. In *The Family Reunion* (1939), one self-described old woman illustrates the futility of temporal action, while her similarly elderly sister presents the redemptive potential of spiritual insight even late in life. The presence of these conflicting concepts in *The Family Reunion*, during the gestation of *Four Quartets*, indicates their significance to Eliot at this point in his life. Amy, the Dowager Lady Monchensey, is the matriarch of her family, and the play is centred upon the titular reunion with her younger sisters, her other sons, relatives and in-laws, to mark the return of her eldest son, Harry, who is to succeed Amy as head of the house. The term 'house' has the dual meaning of both a building and a dynasty, a duality similarly present in the houses which rise and fall in 'East Coker'.

The concepts of age, time and succession merge in the character of Amy. She describes herself several times as an old woman, and, alongside her doctor, as "very much the oldest present [...] as we came first, we will go first".[3] This statement ostensibly claims the right in social tradition to lead the procession in to dinner, but it also reflects her attitude towards an order based upon age and succession, and it presages her imminent departure from life. Her doctor believes that only Amy's will is keeping her alive, determined as she is to see her son Harry succeed her as head of the house. She is said to be "indomitable", a term used to describe unyielding natures predominantly in older people, and an attribution earned in seniority. One character will even suggest that the strength of Amy's will is such that it could have caused another's death. Force of personality is a characteristic uncommon in Eliot's mostly passive elderly characters, who often lack control over life and are either, in his earlier creations, driven by uncontrollable physical urges or, like Gerontion, the Magus and Simeon, in stasis and awaiting release. Against the passage of time, however, even Amy's indomitable will cannot prevail.

Amy believes that, while awaiting Harry's accession, she can halt any change in the house. Upon his return, she declares that nothing has changed, and other characters confirm that the physical characteristics and contents of the house have not altered since his departure. This extends to its staff; the cook is too old to know what she is doing, implying senility, while Hawkins, an elderly gardener, is described as being "nearly half blind".[4] This is not the frequent full blindness of Eliot's other old men, but nevertheless sufficient for the aunts to suggest that Hawkins should be pensioned off. Like Amy, he needs a successor, as Amy has been advised by her sisters "time and time again".[5]

But time, and succession, are precisely what the agèd Amy wishes to control. Succession is time measured in lifetimes; by retaining the elderly staff, and ignoring their physical decline along with her own, Amy is attempting to halt the passage of time and the change it would wreak. Her philosophy echoes the words of the lesser doxology, or *Gloria Patri*, used in Christian traditions at the close of the psalmody, which glorifies the disposition "as it was in the beginning, is now, and ever shall be, world without end".[6] Amy's old age is, in itself, like all old age, a victory over time. In her opening speech, she recalls that, in her youth, she trusted in clocks to mark and ensure the progress and continuity both of time and of life. When she dies at the end of the play, she cries out in her final words, "The clock has stopped in the dark!"[7] Her lengthy life is for her one in a succession of lives; time for her is personal, and the cessation of life is a cessation of time.

With the perspective of youth, Harry challenges his elderly mother's "arresting of the normal change of things".[8] He regards the suppression of change as typical of Amy, but it is typical of old people more widely. Gerontological research into behavioural change, generally focussed upon workplace behaviours and resistance to health advice, has identified factors in older adults which "obstruct attempts at change" and "interfere with the maintenance of change".[9] Harry links the forces of time and change in a single statement: "Time and time and time, and change, no change!",[10] and regards his mother's behaviour as simply "unnatural".

Harry has already declared that "everything is irrevocable, / The past unredeemable",[11] echoing the unredeemable time of 'Burnt Norton'. At the end of the play, Amy accepts this:

> At my age, I only just begin to apprehend the truth
> About things too late to mend: and that is to be old.
> Nevertheless, I am glad if I can come to know them.[12]

This apprehension precedes that gifted by the compound ghost in 'Little Gidding', but there acknowledgement of the past brings shame, pain, and a loss of honour and self-respect in old age. Eliot is clearly aware of the significance of geriatric reconciliation to the past, but Amy's insight by comparison to his later exploration appears shallow and her satisfaction complacent.

Amy identifies with the house, both dynasty and property. When Harry departs at the end of the play, and the succession collapses, Amy's will is broken and her despair expressed through the metaphor of the house itself:

> An old woman alone in a damned house.
> I will let the walls crumble. Why should I worry
> To keep the tiles on the roof, combat the endless weather,
> Resist the wind?[13]

The metaphor recalls Gerontion's decayed and draughty house, a similar image combining physical deterioration in both property and body, but with the added significance here of dynastic failure. "It is no concern of the body in the tomb / To bother about the upkeep",[14] Amy observes of her house. Her will has been broken, and the wind which she no longer wishes to resist is, like the winds of Gerontion and Simeon, the death which she now accepts.

Earlier in the play it is said of Amy's spiritually aware sister Agatha that only she is able to see a meaning in death, and so when Amy's actions fail, and Harry declares that his abdication heralds an end, Agatha responds that it is also a beginning, for she is able to link death to spiritual rebirth:

> I feel very tired, as only the old feel.
> The young feel tired at the end of an action—
> The old, at the beginning.[15]

Agatha presents the prospect of the elderly earning "spiritual income" afresh, despite emotional and physical weariness and a shortage of remaining time.

Amy and Agatha are Eliot's most realised elderly female characters, and the text does not contain any derogatory physical characteristics in relation to them. Their old age is a personification of passing time, and illustrates the futility of attempting to control that passage, while countering the inevitability of physical decline with the possibility of spiritual rebirth even late in life.

In both its title and its content, *Old Possum's Book of Practical Cats* (1939) provides several significant counterpoints to the treatment of old age encountered elsewhere in Eliot's canon. There are many reports that Ezra Pound gave Eliot the nickname 'Possum'.[16] It was one which Eliot frequently employed to sign off letters after 1925, with many variations including Tom Possum, TP and Uncle Possum; when Eliot returned to Harvard in the 1932–33 academic year, and stayed in Eliot House, Pound addressed a letter to Rev Possum, at Possum House,[17] confirming the synonymity of Possum and Eliot. But the adjective 'Old' only began to prefix 'Possum' in Eliot's letters after the first cat poem ('The Naming of Cats') was written for the young Tom Faber in 1936. With variations such as "O. Possum" and "O'Possum", "Old Possum" subsequently replaced most of the versions of the preceding decade; in 1937, a verse sent to Virginia Woolf asks "So what can poor Old Possum do"[18] and he signs off a 1937 letter to Pound as "TP/Old Possum & his Performin [*sic*] Practical Cats".[19] Eliot himself had added 'Old' to his given nickname, 'Possum'.

Eliot's feelings of ageing may have played a part in his choice of epithet. In a 1937 letter to the young Polly Tandy concerning a confusion of dates, he writes:

> Now then ma'am I know that the Old Possum is going motheaten in the fur and shaky in the toofs and uncertain the claws and sometimes he can't even hold on to a branch with his Tail: but I take it hard, I think it crool of you to reb it into him that he is going wambling in the memory.[20]

'Old Possum' is described by some critics as a pseudonym, as if employed by Eliot to distance himself from poems which might have been regarded in literary circles as trivial. M.C. Bradbrook said that the book was "published under an intimate pseudonym",[21] while Dorothy Dodge Robbins, writing in *Names: A Journal of Onomastics*, described 'Old Possum' as "his personal pseudonym".[22] 'Old Possum' and 'Possum' have been included by Carol Yang[23] and Richard Badenhausen[24] in lists of the pseudonyms which Eliot *did* employ to preserve his anonymity, such as Gus Krutzsch and Metoikos. But to describe 'Old Possum' as a pseudonym is inappropriate. Eliot was determined that the work not be seen as "a TSE curio", but felt that it would be unfair on its publishers to publish it anonymously.[25] Consequently, there was no attempt made to hide his authorship of the work. T.S. Eliot is openly credited as the author in several places on the book, from its First Edition onwards (e.g. prior to the positive reception it received after publication). Neither he nor the publisher ever decried the work; the First Edition dustjacket text refers respectfully to "children's verses by Mr. Eliot".[26]

The authorial persona is, however, distinct from the poet himself. The cat poems are not Eliot's, they are Old Possum's; it is Old Possum's 'book'. 'Old Possum' is neither, as we have observed, a pseudonym, nor is it his personal uncompounded nickname of Possum (which Ezra Pound continued to use, referring to Eliot even after his death as "my friend, 'the Possum'".[27]) The *Book of Practical Cats* might be considered closer to a heteronymic work, distinguished from pseudonymic by the Portuguese author Fernando Pessoa, who wrote works in the 'character' of several different authors:

> A pseudonymic work is, except for the name with which it is signed, the work of an author writing as himself; a heteronymic work is by an author writing outside his own personality: it is the work of a complete individuality made up by him, just as the utterances of some character in a drama would be.[28]

Yet the creation of the cat poems in personal letters to children; the technical similarity of his cat poems to his obscene Bolo poems and others; and his stated authorship on the book itself; all mitigate against the idea of 'Old Possum' as "a complete individuality made up by him". Rather, the character of 'Old Possum' permitted the notoriously impersonal Eliot to release in print an aspect of his own personality—an avuncularity which he exhibited and enjoyed in private settings—which runs counter to his carefully established public and social formality. It presents a warm, sympathetic and positive characteristic of old age which, despite being evidently within his own experience, is otherwise absent from his portrayals.

In letters between themselves, Eliot and Pound often adopted the phonetic representation of Southern US Black dialect used in *The Stories of Uncle Remus* (1881) by Joel Chandler Harris, and Eliot would read the Uncle Remus stories aloud to the children of friends and colleagues in the UK, reverting to the accent of his own Southern upbringing. He warned Janet Adam Smith, to whose children he sometimes read, that "As I warm to the theme I tend to lapse completely into my native patois".[29] Contemporary critics have understandably challenged any use by Eliot and others of linguistic imitation and "racial ventriloquism".[30] However, in publishing his animal stories for children as 'Old Possum', Eliot echoes the persona of Uncle Remus, but does not imitate his language. It is the old age of Uncle Remus and his relationship to a juvenile audience which Eliot seeks to adopt.

Uncle Remus is consistently described, and delivers his stories in Harris's books, as "the old man". He claims it is his age which grants him the ability to understand animals; as he expresses it: "w'en you git ole ez me—w'en you see w'at I sees, en year w'at I years—de creeturs dat you can't talk wid'll be mighty skase—dey will dat".[31] By declaring himself to be 'Old', Eliot's possum persona can share this ethological ability. Uncle Remus describes himself here as being "ole". He frequently uses the epithet in descriptions such as "ole Brer Rabbit", but in those it conveys familiarity rather than age. Eliot uses the more formal 'old' to emphasise age and seniority; Old Possum's formality of grammar and outdated diction—"on the occasion of which I shall speak"[32]—is that of an elderly narrator.

Joel Chandler Harris describes the affectionate relationship between Uncle Remus and the child, who persuades the old man to tell his stories: "The little boy [...] was in thorough sympathy with all the whims and humors of the old man, and his capacity for enjoying [the stories] was large enough to include even those he could not understand".[33] Eliot was seen to pursue a similar relationship himself in reading to children, as Janet Adam Smith observed:

> Like Uncle Remus with the little boy Possum had to be coaxed. Did they really want it? Book pressed into his hand, with some merriment; affected surprise; adjustment of spectacles; much turning to and fro of pages; renewed reluctance; increasing encouragement from audience – then [...] we were well away for another hour.[34]

Clearly mirroring Eliot's own relationship with children, the character of Old Possum presents avuncularity as a positive characteristic of old age. It is the first time we have seen an elderly character in Eliot's work who inspires affection. His tone evinces what Harris describes as "the air of affectionate superiority which Uncle Remus assumes"[35]; Eliot achieves this through frequent iterations of pedagogic phrases—"as I said", "your memory I'll jog", "I tell you once, and once again".

In addition, Old Possum suggests that he is sharing privileged information accrued in his seniority—"I am tempted to say", "I have known"—which welcomes the audience into his superior status. Placing an old man in a position of authority over the reader echoes Tiresias in *The Waste*

Land, whose knowledge informs his observations; but Old Possum shares his privileged information, while in his historical sources Tiresias often did not.

Four of the cats named and characterised within the final collection of Practical Cats are old. Two are titled as such ('The Old Gumbie Cat' and 'Old Deuteronomy'), and two are revealed as such ('Gus the Theatre Cat' and 'Cat Morgan'). Creatures which are clearly intended to inspire affection are given old age among their other characteristics, an attribute which might have engendered in Eliot's earlier work negative emotions ranging from pity to revulsion. Perhaps by receiving from an audience of children an affirmative, affectionate response to aspects of old age, Eliot was able to acknowledge its positive qualities.

'The Old Gumbie Cat' evinces two of the positive geriatric qualities which he has previously acknowledged in old men. She is contemplative; she "sits and sits and sits and sits", and this is fundamental to her character, as it is "what makes her".[36] This is the contemplative condition to which Eliot aspires in 'Ash-Wednesday': "Teach us to sit still".[37] But unlike many of Eliot's old men, she is also capable of action, in equal measure to her contemplation; and when she acts, she is in control. She imposes order upon the mice and the "disorderly" cockroaches. She is responsible for "well-ordered households", an imposition of authority by an elderly character which reflects Eliot's critical association of maturity and order (see Chapter 4).

In a letter to seven-year old Polly Tandy, Eliot explains that Old Deuteronomy is "a Very Old Cat; because he has only forty or fifty years more to live, and I wanted him to have the glory of a Poem about him now while he could appreciate it". He explains that Deuteronomy means 'second name', and that this replaced his younger name, which was quite ordinary, "when he became quite old – I mean about forty-eight or nine".[38] This was Eliot's own age at the time. It is stated that Old Deuteronomy's "fame" was garnered prior to the 1837 succession of Queen Victoria, which would actually render him greater than centenarian. He has "lived many lives", and has "numerous progeny", suggesting a full and active life. Although now "in his decline", Eliot portrays that as a distinguished decline; Old Deuteronomy is "placid and bland", contrasting with the village's (human) "Oldest Inhabitant" whose croaking voice, wandering mind and tottery legs serve to emphasise the cat's serenity. Old Deuteronomy is present at three key village locations:

the church, the market and the pub, representing respectively the spiritual, economic and social centres of life. These three are not in conflict, and in each environment he is accorded respect, a Geront who is acknowledged and supported, and whose earned repose in the later years of life is ensured by the community.[39] He demonstrates in old age the positive renunciation of action which underpins 'A Song for Simeon'.

The physical decline of Gus the Theatre Cat in his old age is more vividly portrayed—"he isn't the Cat that he was in his prime". He is described as thin, and suffering from palsy, and although smart in his youth, he is now shabby in appearance. "He once was a Star of the highest degree"; but his fall from that position has clearly been long and slow, since of those with whom he acted, (Herbert Beerbohm) Tree died in 1917, while (Henry) Irving died in 1905, thirty-four years before publication. There is a suggestion that despite his theatrical career, he once had the kind of aggression praised in some of the other, younger cats, but he is "no longer a terror" to rats and mice.

Nevertheless, his decline is again treated affectionately. He is not alone, but has friends, whom he regales with his reminiscences. Furthermore, instead of the bitterness or regret commonly suffered by Eliot's old men considering their past, he speaks with pride of his "palmiest days", his skills and abilities, his triumphs and success.[40] History for Gus is not cunning or contrived, but straightforward; he has established his legacy, and it will tell of his grandest dramatic creation.

Eliot wondered whether the character of Gus might be too "antiquated"[41] for his audience, after his friend Geoffrey Tandy, Polly's father, questioned the relevance to young people, who were eschewing theatregoing, of "the figure of the broken-down actor [...] living on his past triumphs". However, acting offers examples of a failing career in the public eye, a waning which remains unseen in other, less public professions. Gus's retention of his dignity in the face of such a socially visible decline is another supportive perspective on old age. Tandy also suggested to Eliot that "perhaps there's something to be said for the poor old has-beens, as it might be you or me".[42]

Unlike the other Practical Cats, Morgan was based on an actual cat, a stray which added itself to Faber's offices, and provided Eliot with comfort during nights of fire-watching at Faber during the War. The poem was written in 1951[43] and used as Faber publicity; Morgan, described as "venerable" by Eliot when he died "of extreme old age" in July 1952, was added to the original collection in 1953.

Morgan was once a pirate, a role embodying the "terror" which several of the cats have engendered. He is now "Retired"—later unpublished lines written for a Faber-related birthday describe him as "too old for fightin'"[44]—and he has suffered some physical decline; he "got knocked about" as a pirate. But he is popular ("You can't but like Morgan"), has a good heart, is smart in appearance, even sexually attractive, and despite being rough-spoken he commands respect from visiting businessmen. His food and drink leave him "content", and he describes himself confidently as "a-takin' my ease".[45] It appears to be a quiet and dignified 'retirement' at Faber & Faber, something which Eliot himself at sixty-three might have found appealing.

The old age of Old Possum's cats present striking contrasts to the old men of Eliot's poetry. Old Deuteronomy's position in the three key arenas of life—spiritual, economic and social—engenders respect, where Eliot's old men frequently encounter conflict. The Old Gumbie Cat can successfully combine both contemplation and action. Gus retains his dignity and pride in the face of career decline; and Morgan is popular and relaxed in his retirement role. Presented by an avuncular old figure who inspires affection, all exhibit contentment in their dotage. In the face of his other negative representations, *Old Possum's Book of Practical Cats* is an acknowledgement by Eliot of commonplace satisfactions in senescence, attributing to his cats and to his Old Possum persona many positive traits of old age which he never attributes to his human characters.

It is in *Four Quartets* that Eliot finally addresses his personal experience of old age. The work's absence of personae coincides with the absence of the established and predominantly negative geriatric imagery which Eliot 'left behind' in 'Ash-Wednesday'. Instead, he speaks directly, in the conversational tone which occurs frequently in the *Quartets*, and identifies, in a manner sometimes critical, sometimes self-accusatory, a litany of failings and 'wrongs' in old men, including impotent rage, failing pride, declining powers, expiring senses, shame, exasperation, late awareness and a fear of belonging.

Over the eight years from the composition of 'Burnt Norton' in 1935 to the publication of the completed *Four Quartets* in 1943, old age becomes an increasing presence in the work, culminating in the emotional revelations concerning old age by the compound ghost in 'Little Gidding'. Whilst the exploration of time within 'Burnt Norton' (1936), which opens *Four Quartets*, concentrates upon detachment from the passage of human time, and therefore from age, Eliot nevertheless describes "the weakness of the changing body", acknowledging the presence of physical ageing. The observation in the opening passage that "All time is unredeemable" seems an abstraction, but over the course of the subsequent Quartets it will become a profoundly personal acknowledgement, as the poet reaches an age which imposes the "rending pain of re-enactment" of actions and motives from the past.

'East Coker' (1940) is the first of the *Quartets* to examine the actuality of the old age which Eliot had long anticipated. He perceives himself now to be in the autumn, the "late November", of his life. Yet he also describes himself as

> In the middle, not only in the middle of the way
> But all the way, in a dark wood, in a bramble,[46]

The lines echo the opening of Dante's *Inferno*: "In the middle of the journey of our life I came to myself in a dark wood [...] a wild, and rough, and stubborn wood" (in Eliot's favoured Temple Classics translation). Dante is assumed to be writing aged thirty-five, in the middle of his Biblical lifespan of 'three score years and ten'. Eliot, writing aged fifty-one, might still be considered objectively middle-aged; but we have seen his persistent feelings of old age, and the concern he expressed about his remaining years of activity, perhaps reflected here when he insists that he is not just in the middle, but *all* the way. When he praises Yeats for declaring in *Responsibilities* his age as forty-nine, Eliot describes that as "More than half a lifetime".[47]

Eliot also considered himself to be in the middle of three generations of poets. In a letter of 1938, he said:

> It is perhaps useful to remember that there are three generations of poets alive, each with its own characteristics. The oldest is Mr Yeats; the middle generation may be represented by Mr Ezra Pound and myself; and the third by Mr Auden and Mr Spender.[48]

Yeats died prior to the publication of 'East Coker', but the three generations can nevertheless be regarded as a sequence of poetic significance across the two World Wars, retaining Eliot's position 'in the middle' with Pound, in what is effectively the middle of the twentieth century.

Eliot now questions the anticipated benefits of age, which were "long looked forward to, / Long hoped for".[49] His use of 'long' acknowledges the years through which he has considered old age, and questioning the absence of those benefits confirms that he considers himself now to be of that age when they should be granted. But old age is not as he had expected. He had anticipated calm, serenity and wisdom, although these are not characteristics which have been present in his many portrayals of old men; indeed, it has often been wisdom which has disrupted the calm and serenity of his characters' final years. However, 'wisdom' is now seen as only the possession of useless secrets from the past, while 'serenity' is a "deliberate hebetude" recalling Gerontion's lethargy. Indeed, as Kathleen Woodward observes,

> Although only middle-aged — and this is still a time of building — Eliot portrays himself as paralyzed, particularly in terms of poetry and the future. At the same time, he dismisses the tranquility of the wise old man for himself — and for others. "Autumnal serenity," we read in the second section of 'East Coker', is in fact lethargy, dullness of mind, a reliance on old insights and useless knowledge.[50]

The absence of these benefits Eliot blames upon deception by "elders", the term referring not to those presently older than Eliot himself but to figures now dead, whose words have been "bequeathed" to him. Eliot believes that those elders may have been deceiving themselves about the quality of their old age; Eliot now regards their supposed calm and serenity as dullness and lethargy, and their wisdom as useless. In particular, he questions the value of the wisdom which is commonly regarded as the most significant positive attribute of old age—the benefit of accrued understanding, or as Eliot describes it, "knowledge derived from experience".[51]

The use of "elder"', and of the older term "receipt" for the recipe which they bequeathed, refer back to Thomas Elyot, Eliot's ancestor who wrote *The Boke named The Governour* (1531), a treatise quoted in the first section of 'East Coker', setting out a way of life for the governing class in England. Because many of the sixteenth-century landed class lacked educated learning, it was expedient for Elyot to praise "EXPERIENCE whereof commeth wysedome". Learning, as he described it, was only the acquisition "Of Experience whiche haue preceded our tyme".[52]

> No operation or affaire may be perfecte, nor no science or arte may be complete, except experience be there unto added, whereby knowlege is ratified, and (as I mought saye) consolidate.[53]

This is the deception Eliot believes that he has been bequeathed. Far from a benefit, Eliot avers that the experience inherent in old age imposes a repetition of those past experiences, taking no account of the present, and oblivious to the concept that each present moment should offer a re-evaluation of our past. This accords with the historical sense which he posited in 'Tradition and the Individual Talent', "a perception, not only of the pastness of the past, but of its presence", and with the belief he expressed there "that the past should be altered by the present as much as the present is directed by the past".[54] The greater one's age, the greater one's experience—not just one's own, as he expands in 'The Dry Salvages', but also (as the historical sense suggests) the experience of preceding generations.

Eliot is able now to see that the past is not simply a sequence, still less a development, but rather a "pattern", a "more complicated" pattern, "Of dead and living", which can only be perceived with the perspective of old age. To dismiss the wisdom handed down by the elders as "useless", Eliot employs his familiar trope of failing elderly eyesight, depicting the old men "peering" into darkness, and turning their eyes away, as their perception fails. Eliot goes on to enumerate further failings in old age. In this "final addition" of life will be reckoned a failing of both pride and powers—not 'power' per se, but creative powers—as he continues to wrestle with language. The continued "trying" must, in itself, be sufficient. He is not adopting the bitterness of the narrator of 'Lines for an Old Man'; rather, it is his pride which will suffer from his awareness that, as he will say in 'Little Gidding', "next year's words await another voice".

Discounting the assumed wisdom of old men, Eliot identifies:

> Their fear of fear and frenzy, their fear of possession,
> Of belonging to another, or to others, or to God.[55]

Just prior to Eliot's composition of 'East Coker' (and his lecture on Yeats), *Last Poems and Two Plays* by Yeats was published in July 1939, a posthumous collection following his death earlier that year. "Grant me an old man's frenzy", Yeats declaims in the collection's 'An Acre of Grass'. Yeats presents 'frenzy' as a positive attribute of old age and, far from fearing it, yearns for “A mind…inspired by frenzy”. His poem ends with a reference to “an old man’s eagle mind”[56] which would have resonated with Eliot’s image of himself as an agèd eagle, while the associations of an eagle’s sharp focus counter any notions that ‘frenzy’ might represent senility. The fear of belonging to another, whether temporally or spiritually, reflects the importance which individuality and a belief in personal autonomy has come to hold since the Renaissance period of Thomas Elyot. The significance of personal identity, to which old men cling in the pattern beyond their experience, is perhaps their greatest folly—and is the antithesis of humility, which is “The only wisdom we can hope to acquire”.

In the drafts of ‘East Coker’, Eliot emended an initially abstract “As one grows older” to the more intimate “As we grow older”,[57] a personal acknowledgement of increasing age. The perspective which this affords does not clarify the past, however, but only exposes its complexity. In old age, the world becomes “stranger”, the term combining in one sense odd and confusing, and in another unknown and alien, both interpretations of their contemporary world which are characteristic of those in their dotage. Old age brings experience into the epiphanic moment which, instead of being isolated from time, now contains the past, the literal “lifetime” of a man. This moment must necessarily increase with old age, as the lifetime itself increases. It comes together with the incomprehensible (because not experienced) historical past of those whose lifetime has ended, in the image of the photograph album.

In a letter to his mother after receiving some family photographs from home, Eliot wrote:

> It gives one a strange feeling that Time is not before and after, but all at once, present and future and all the periods of the past, an album like this”[58]

This property of photographs to capture time was significant to Eliot. His sitting room would display more than a dozen family and related photographs, including relatives, like his grandfather William Greenleaf Eliot, who had died before he was born.[59] In *The Family Reunion*, almost contemporaneous with ‘East Coker’, the concomitant notion of the destruction of the past employs the same image, when “the wings of the future” are said, in their desecration of history, to have “mutilated the family album".[60] The elderly can return to their past when it is held within their photographs.

In 'The Dry Salvages' (1941), "one becomes older", adopting a position of completion after "we grow older" in 'East Coker', and can therefore refer to "years of living". Published only eleven months after 'East Coker', there is little change in the observations concerning old age which it presents. This stasis is in accord with the poem's notion that, as one ages, having gained the perception of the world as strange and the pattern of life as complicated, one must abandon the idea of life as a linear sequence or development.

'The Dry Salvages' reiterates the previous Quartet's notion that the past of previous generations is contained within present experience. However, it adds that in old age, one discovers that moments of agony, including those of others, survive from the past and are similarly permanent. Time will pass, and events may be obscured by subsequent actions, so the "patient" who arrives at old age is not the person who began the journey. Time is no healer.

In 'East Coker', Eliot declares that "Old men ought to be explorers".[61] Eliot commonly ascribes contemplation and action to old age and youth respectively; here, exploration is associated with a departure from home, the 'East Coker' from which Eliot's forebears set sail for America, and hence the succeeding marine imagery leading into 'The Dry Salvages'. However, these old men must be "still and still moving / Into another intensity",[62] physical movement replaced with contemplative exploration, taking their lifetimes into the intense moments of the preceding lines. 'Little Gidding' (1942) will declare that the end of their exploration will be at their beginning, known for the first time—and that their exploration will not cease. Indeed, it is in 'Little Gidding', published when he was fifty-five, that Eliot "disclose[s] the gifts reserved for age". No longer an intellectual projection into an elderly persona, this expression is of a personal, emotional experience of ageing. "More than half a lifetime", as he said of Yeats, "to arrive at this freedom of speech".[63]

Given his persistent use of wind as a portent of death, it is indicative of Eliot's state of mind that the "familiar compound ghost" in 'Little Gidding' arrives "as if blown [...]Before the urban dawn wind".[64] The compound ghost should not be seen as a persona; Eliot "assumes a double part", speaking as both the ghost and himself. As Helen Gardner clarified, "the ghost speaks with Eliot's own voice. It is, in a profound sense, a meeting with himself".[65]

Nevertheless, the compound nature of the ghost combines several identifiable 'past masters' who could speak to Eliot of old age. The encounter draws upon Dante's meeting in *Inferno* with his own former teacher Ser Brunetto, who would have been around seventy-four at the time of Dante's writing. Their conversation explores the prospects of success in life, and whether Dante has lived up to the expectations of his old master. Dante tells Brunetto that "*m'insegnavate come l'uom s'eterna*"—"you taught me how man makes himself eternal".[66] As Eliot assesses his life 's achievements and what he had learnt, employing a poetic form drawn from the *Commedia*, he is aware of how words have made Dante himself eternal. Taking further inspiration from the *Commedia*, the ghost also embodies Virgil, a figure from whom Eliot learnt, and who guided Dante as a seeking poet.

Jonathan Swift is another 'past master' whose values are incorporated into the compound ghost. In a 'Commentary' in *The New Criterion* Eliot quotes approvingly a paraphrased view of Swift voiced by Kipling: "A man of overwhelming intellect and power goes scourged through life between the dread of insanity and the wrath of his own soul warring with a brutal age".[67] It is a description to which Eliot himself might have been pleased to lay claim. Eliot remarked Swift's power of scorn, and perception of folly and stupidity,[68] and praised his "terrible sincerity",[69] all characteristics which will be present in the compound ghost's words concerning old age.

It is however the sincerity and directness which Eliot found in late Yeats which is most pervasive in the ghost's disclosure. In a letter to John Hayward during composition. Eliot acknowledged that while his intentions were not so precise, most readers would identify the compound ghost with Yeats. Hence Eliot's deletion of any specific reference to Brunetto beyond the ghost's "brown baked features", as although he was happy with the link to Dante, he would not wish Yeats to be directly associated with the sin which had condemned Brunetto to the *Inferno*.[70]

The syntax of the compound ghost's opening offers two interpretations; either the gifts, or the disclosure of the gifts, can crown the lifetime's labours. The disclosure of the gifts is a poetic achievement in *terza rima,* meeting the objective, shared between Eliot, his ghost, and their past masters, "to purify the dialect of the tribe" (in the words of Mallarmé), overcoming the wrestle with language which Eliot has lamented in preceding *Quartets*. This passage of disclosure is not only a formal triumph but also a first-person confession of a kind which Eliot has not previously exhibited, providing the "acute personal reminiscence" which he was aware that the draft poem lacked.[71] Together, the form and content of the passage might in themselves suffice to crown his lifetime's efforts.

However, more directly, there are, laden with irony, the actual gifts reserved for old age (or "final prizes" in an earlier draft),[72] bestowing upon senescence the perception of a series of physical, emotional and moral failings. We have seen how the influence of late Yeats impelled Eliot towards emotional expression; what it also gave Eliot was an example of a poet "who, out of intense and personal experience, is able to express a general truth; retaining all the particularity of his experience, to make of it a general symbol".[73] It was an exemplar he sought to emulate by making of his own experience and personal reminiscence not only powerful confessional poetry, but an expression of universal characteristics of old age.

The first attribute disclosed is the "cold friction of expiring sense". This recalls Gerontion, whose "sense has cooled", offering no prospect of physical bliss, only a dysphoric frustration with disappointing echoes of previous fulfillments. As early as his prose fiction *Eeldrop & Appleplex* (1917) Eliot had imagined in old age "all other appetite gone except the insatiable increasing appetite of vanity".[74] A speech by Edward in *The Cocktail Party* links the advent of old age explicitly to the loss of desire:

> That is the worst moment, when you feel that you have lost
> The desire for all that was the most desirable,
> Before you are contented with what you can desire;
> Before you know what is left to be desired;
> And you go on wishing that you could desire
> What desire has left behind.[75]

Previous observations advocated a positive, "inner freedom" from desire.[76] This ascetic 'Saint's way' is not, however, the experience of "most old men". Edward's circumlocutory speech, in a drama featuring relationships between characters of disparate ages, confirms the expiring of sense presented in this first gift as the "worst moment" of ageing.

We are told by the ghost that last season's fruit has been eaten. As a symbol of past literary work, both his own and in his borrowings, the phrase resonates with Eliot's earlier overtly allusive technique, increasingly absent from his later work. The sweetness of that fruit is now decayed; an old man is left instead with tasteless "shadow fruit", the shadow not only suggesting achievements both illusory and evanescent, but also carrying the sinister connotations with which Eliot often imbues the term, as in the threatening shadows in *The Waste Land*, the shadow that led men to darkness in *The Rock*[77] and the shadow which falls between opposed concepts in 'The Hollow Men'.[78] The taste, the satisfaction, the sensory sweetness is absent from the fruits of labour, as body and soul grow "asunder', separating as well as deteriorating in old age.

The second 'gift' is senescent rage at human folly. For the Magus, folly was an accusation to be ignored, overcome by faith; in 'East Coker' folly lay in the mistaken perception of old men's wisdom. Harry in *The Family Reunion* declares folly to be "The instinct to return to the point of departure / And start again as if nothing had happened"[79] a near-aphoristic definition of folly as repetition driven by the past, at which Eliot himself rages in 'East Coker'. Here folly is the cause of geriatric rage, described as not only impotent, but consciously impotent, an old man aware of the emotion's futility but unable nevertheless to suppress it. The rational mind, exhibited by Eliot throughout his life, is surrendered to the power of senescent emotion.

The final 'gift' transforms geriatric reminiscence, which Eliot has used as a poetic device since his graduation address, into a painful act of critical self-awareness. In *After Strange Gods*, acknowledging that Yeats "has at least discarded, for the most part, the trifling and eccentric"[80] of which Eliot had been so critical, he expresses admiration for Yeats's austerity, and the regret and resignation voiced within his later poetry. He quotes from Yeats's 'Vacillation':

> Things said or done long years ago,
> Or things I did not do or say
> But thought that I might say or do,
> Weigh me down, and not a day
> But something is recalled,
> My conscience or my vanity appalled.[81]

These lines from Yeats resonate with the ghost's final disclosure. The recollection of, and acceptance of responsibility for, past errant behaviour is fundamental to both. Eliot's contention that Yeats's observations apply to most other old men raise his own parallel observations from the purely personal to the universal.

Old age is not, indeed, what one had expected. In his graduation poem of 1905, the teenage Eliot projected himself into old age in order to look back on the past. Upon reaching that imagined old age, projection has been replaced by retrospection, fond reminiscence has been replaced by painful re-enactment, and honour has been stained. Things not done, which defined Gerontion, have been surpassed in significance by things ill-done. The self-criticism expressed here echoes the reason Eliot gives Emily Hale for wishing his letters to be kept secret; rather than the exposure of romantic intimacies, he is "ashamed of [...] the exhibition of vanity, boastfulness, peevishness, perhaps even malice unconscious – and of all the callow and mistaken judgments upon people that one has made in the past".[82]

The abstract concept with which *Four Quartets* opens, that all time is both eternally present and unredeemable, has become an emotional personal experience; Eliot questions past actions and disinters motives and consequences, but when that past cannot be redeemed, it offers only exasperation and shame. The fact that this was unforeseen in youth enhances the impact of its disclosure. What might have been remains an abstraction, but what was once an intellectual conceit has become in old age a source of emotional pain.

Eliot's self-declared old age was the cause of surprisingly personal comments from some who found poetic failings in *Four Quartets*. Karl Shapiro noted "a kind of narcissistic figure of the aging Eliot lolling through the poem".[83] David Perkins remarks upon Eliot's need "to find... a reconciliation to life in time, a need especially urgent since the protagonist is himself an old man who is learning that 'time is no healer'".[84] And in 1949, Stefan Schimanski and Henry Treece introduced *A New Romantic Anthology*, intended, counter to Eliot's own philosophy, "to show how widely spread is the Romantic attitude in British writing today". Declaring that "One may only find truth by exploration", an obvious echo of Eliot's "old men ought to be explorers", they concluded with an *ad hominem* dismissal of "the Classical mind, or what we would call the staid and restricted personality, for whom life's adventure has ceased, has no time, no necessity, and no energy for such a freedom of search".[85] Of those accusations, only the absence of necessity stands as a valid reason for Eliot producing no major poetry after *Four Quartets*. That work, wrought from his old age, amply crowns his lifetime's effort in poetry. As he proceeds into later life, however, old age will continue to announce its presence in his creative, public and private life.

Notes

1. Sencourt, Robert. 1971. *T.S. Eliot: A Memoir*. Garnstone Press, 146.
2. Eliot, T. S. 2022. *The Complete Prose of T. S. Eliot: The Critical Edition Vols 1–8*. Ed. Schuchard, Ronald, Brooker, Jewel Spears, Cuda, Anthony, Dickey, Frances, Formichelli, Jennifer, Harding, Jason, Javadi, Iman, Stayer, Jayme and Chinitz, David. The Johns Hopkins University Press and Faber & Faber Ltd. http://muse.jhu.edu/resource/1 4, 373.
3. Eliot, T.S. 1939. *The Family Reunion*. Faber & Faber, Part I, Sc III.
4. Eliot, *The Family Reunion* Part I Sc I.
5. Ibid.
6. *Directorium Anglicanum*. 1858. 97, 134.
7. Eliot, *The Family Reunion*. Part II Sc III.
8. Eliot, *The Family Reunion* Part II Sc II.
9. National Research Council (US) Committee on Aging Frontiers in Social Psychology, Personality, and Adult Developmental Psychology; Carstensen LL, Hartel CR, editors. *When I'm 64*. Washington (DC): National Academies Press (US); 2006. 3, Motivation and Behavioral Change: https://www.ncbi.nlm.nih.gov/books/NBK83771/.
10. Eliot, *The Family Reunion* Part I Sc I.
11. Eliot, *The Family Reunion* Part I Sc III.
12. Eliot, *The Family Reunion* Part II Sc III.
13. Eliot, *The Family Reunion* Part II Sc III.
14. Eliot, *The Family Reunion* Part II Sc III.
15. Eliot, *The Family Reunion* Part II Sc II.
16. Eliot, T.S. 2015. *The Poems of T.S. Eliot Vols I and II 1.* Ed. Christopher Ricks and Jim McCue. Faber and Faber 2, 37.
17. Levin, Harry. 1985. 'Old Possum at Possum House', *Southern Review*, 21 (October 1985), 1008.
18. Eliot, T.S. 2009–2025. *The Letters of T.S. Eliot Vols 1–10* Ed. Valerie Eliot, Hugh Haughton and John Haffenden. Faber & Faber 8, 675.
19. *Letters* 8, 742-3.
20. *Letters* 8, 605.

21. Bradbrook, M.C. 1965. *T.S. Eliot*. University of Nebraska Press, 64.
22. Robbins, Dorothy Dodge. 2013. 'Imperial Names for "Practical Cats": Establishing a Distinctly British Pride in Old Possum's Book of Practical Cats', *Names: A Journal of Onomastics*, Vol. 61 No. 1, March 2013, 23.
23. Carol L Yang. 2012. 'T.S. Eliot's virtual Europe: the Flaneur and the textual Flanerie',
    Yeats Eliot Review (Vol. 29, Issue 3–4), 7.
24. Badenhausen, Richard. 2004. *T.S. Eliot and the art of collaboration*. University Press, Cambridge, 29.
25. Eliot, *Letters* 9, 211.
26. Eliot, T.S. 1939. *Old Possum's Book of Practical Cats*. Faber & Faber.
27. Pound, Ezra. 1966. 'For T.S.E.', *The Sewanee Review*, Winter 1966, Vol LXXIV, No 1, 109.
28. Pessoa, Fernando. 1995. 'Tábua Bibliográfica in *Presença* 17, December 1928', trans. *A Centenary Pessoa*, ed. Eugénio Lisboa with L.C. Taylor, Carcanet Press, 133/4.
29. Adam Smith, Janet. 1985. 'Tom Possum and the Roberts Family', *The Southern Review*; Oct 1, 1985, Autumn, 21, 4.
30. North, Michael. 1998. *The Dialect of Modernism: Race, Language, and Twentieth-Century Literature*. Oxford University Press I, 1.
31. Harris, Joel Chandler. 1881. 'Mr Fox and the Deceitful Frogs', *Uncle Remus his songs and his sayings: the folk-lore of the old plantation*. D. Appleton and Company, XIV, 68.
32. Eliot, *Poems* 2, 18.
33. Harris, *Uncle Remus*,'The End of Mr Bear', XXVIII, 120.
34. Adam Smith, 'Tom Possum', 1065.
35. Harris, *Uncle Remus*, 12.
36. Eliot, *Poems* 2, 6.
37. Eliot, *Poems* 1, 97.
38. Eliot, *Letters* 8, 709.
39. Eliot, *Poems* 2, 16.
40. Eliot, *Poems* 2, 25.
41. Eliot, *Letters* 8, 792.
42. Eliot, *Letters* 8, n792.
43. Eliot, *Poems* 2, 75.
44. Eliot, *Poems* I, 308.

45. Eliot, *Poems* 2, 35.
46. Eliot, *Poems* 1, 187 Later in 'East Coker' Eliot describes himself as "in the middle way, having had twenty years [...] Trying to learn to use words" (Eliot, *Poems* 1, 191), a period which accords with the time since his first published collections.
47. Eliot, *Prose* 6, 81.
48. Eliot, *Letters* 8, 782.
49. Eliot, *Poems* 1, 187.
50. Woodward, Kathleen M. 1980. *At Last, the Real Distinguished Thing: the late poems of Eliot, Pound, Stevens, and Williams.* Ohio University Press, 58.
51. Eliot, *Poems* 1, 187.
52. Elyot, Thomas. 1962. *The Book named The Governor*. J.M. Dent & Sons, Second Everyman's Library Edition, III, xxv, 228.
53. Ibid., III, xxvi, 231.
54. Eliot, *Prose* 2, 106.
55. Eliot, *Poems* 1, 188.
56. Yeats, W.B 1983, *The Poems*, ed. Richard J Finneran, Macmillan, 301.
57. Gardner, Helen. 1978. *The Composition of Four Quartets.* Faber and Faber, 110.
58. Eliot, *Letters* 1, 243.
59. Eliot, *Letters* 3, 647.
60. Eliot, *The Family Reunion*, Part 1, Sc III.
61. Eliot, *Poems* 1, 191.
62. Eliot, *Poems* 1, 192.
63. Eliot, *Prose* 6, 81.
64. Eliot, *Poems* 1, 203.
65. Gardner, *Composition*, 185.
66. Dante, *Inferno*, trans.Temple Classics, J.M. Dent & Sons, Canto XV, 161.
67. Eliot, *Prose* 2, 830.
68. Eliot, *Prose* 3, 451.
69. Eliot, *Prose* 4, 373.
70. Eliot, *Letters* 10, 183.
71. Eliot, *Letters* 9, 884.
72. Gardner, *Composition*, 190.
73. Eliot, *Prose* 6, 81.
74. Eliot, *Prose* 1, 529.

75. Eliot, T.S. 1950. *The Cocktail Party*. Faber & Faber, Act 1, Sc II.
76. Eliot, *Poems* 1, 181.
77. Eliot, *Poems* 1, 168.
78. Eliot, *Poems* I, 83–4.
79. Eliot, T.S., *The Family Reunion*, Pt I, Sc II.
80. Eliot, *Prose 5*, 36.
81. Yeats, W.B. *The Poems*, 251.
82. Eliot, T.S. Letter to Emily Hale, 27th October 1956.
83. Shapiro, Karl. 1960. *In Defense of Ignorance*. Random House, 56.
84. Perkins, David. 1962. 'Rose Garden to Midwinter Spring: Achieved Faith in the Four Quartets', *Modern Language Quarterly* 23 (1): 41–45 https://doi.org/10.1215/00267929-23-1-41.
85. Schimanski, Stefan and Henry Treece, Henry. 1949. 'Introduction: The Personalist view of Romanticism', *A New Romantic Anthology*. The Grey Walls Press, 19-20.

CHAPTER 8

"An Elder Statesman": 1947–1965

Abstract In his own old age, Eliot's literary and public status called for him to write many tributes, eulogies and memorials for passing members of his generation; they reflect not only his critical judgements upon others' lifetimes of work, but also an abandonment of impersonality, and a desire in his old age to confirm his personal legacy. His final Ariel poem returns to the poetic device first employed as a teenager, with the reminiscences of an old man; while his final play, *The Elder Statesman*, presents a titular character reflecting not only Eliot's public persona, but his personal concerns about his past and his late realisation of the power of human love. In 'A Dedication to My Wife' he celebrates his unexpected discovery of love and sensual passion in old age, repudiating his lifelong vision of senescence, and concluding the old age which he had presented with such negativity in unanticipated satisfaction.

Keywords Elder Statesman · Cultivation of Christmas Trees · Authored obituaries · Dedication to my Wife

Eliot's first wife, Vivienne, from whom he had been formally separated in 1933, died in January 1947, in the mental hospital to which she had been committed in 1938. Eliot wrote to Emily Hale of the moment when her coffin was settled in the ground: "Suddenly I felt as if I had

P. Keers, *T.S. Eliot*, Modern and Contemporary Poetry and Poetics,
https://doi.org/10.1007/978-3-032-25050-6_8

been preserved in a mummified youth, and at that moment I became my chronological age – not a young man to whom opportunities are open, but a senior who had lived his life, such as it was, a man getting on for fifty-nine, with limited possibilities for the remaining years".[1]

The old age which Eliot had anticipated all his life descended upon him physically in the 1950s. Meeting him for the first time, the American poet Donald Hall observed that "Eliot was only sixty-three, in the autumn of 1951, but he looked at least seventy-five". Hall describes a pale face, stoop and hacking smoker's cough. "Eliot", he concludes, "was cadaverous".[2] Eliot spoke to Mary Trevelyan in the following year of his dread of old age, with its difficulties and complications[3]; hospitalised variously with tachycardia, emphysema and bronchitis, he reconfirmed his own funeral arrangements.[4]

On the mantelpiece of his Faber office Eliot displayed a facsimile of a print by Daniel Maclise of Goethe in old age.[5] Lecturing on Goethe in 1955, Eliot described its image of a weakened body still ruled by a vigorous mind, of a man who combined youthful vitality with the wisdom of age. For Eliot, the fact that Goethe could combine that wisdom with "poetic speech" confirmed his status as a great poet.[6] It was clearly an old age to which Eliot aspired.

He had fulminated, in his essay on Yeats, upon elderly poets "becoming dignified, of becoming public figures with only a public existence – coat-racks hung with decorations and distinctions".[7] But he had also written to Emily Hale, upon being awarded an honorary degree:

> It is interesting, as one gets older, that one is pleased – tickled in fact – by trifles of this sort which would have seemed indifferent and outrageous in youth. Perhaps it is that in age one has learnt to be content, in this world, with trifles.[8]

Eliot defended his "outside activities" in a letter to John Hayward of March 1944. "I am sure that when I am eventually weighed in the balance", he wrote, "it will only be the poetry (such as it is) which will tip the beam in my favour: everything else will be feathers". But he explained that as for other activities,

> there is partly the need for new aliment for poetry – perpetual transformation of thinking and feeling; partly the need to distract my consciousness

> from poetry, so as to let it work under the surface, and never force it – partly the inherited Social Conscience and the desire to show (to myself) that I can count for something in the kind of activities that my family have pursued for centuries; and partly, of course, the normal sediment of vanity and sloth (it is sloth that makes one do a great deal of work).[9]

While Eliot continued to write plays, having been awarded the ultimate 'decorations and distinctions' with the Nobel Prize and the Order of Merit, he appeared across post-War Europe and America on behalf of the British Council as a cultural ambassador. "The years between 50 and 70 are the hardest", he told an interviewer. "You are always being asked to do things, and yet you are not decrepit enough to turn them down".[10]

Eliot wrote numerous public letters, particularly to *The Times*, both under his own name and as a co-signatory, on matters literary, clerical and social. And as a senior member of the literary establishment, he wrote a number of tributes, obituaries and memorials. "I certainly hope that I do not see myself as a Public Figure!"[11] he was quick to retort to Emily Hale. But as a character observes in his final play, "The difference between being an elder statesman and posing successfully as an elder statesman is practically negligible".[12]

When Faber decided to revive its series of Ariel Poems, 'The Cultivation of Christmas Trees' (1954) was Eliot's contribution. In the letters to Emily Hale following the death of Vivienne in 1947, Eliot anticipated that he had between ten and fifteen years of activity remaining.[13] When 'The Cultivation of Christmas Trees' is published, he has just turned sixty-six and reached the mid-point of that prognostication. But the poem itself conflates his eightieth Christmas with his last, echoing Simeon's "eighty years and no to-morrow"[14] as a lifespan, and anticipating a further fourteen years of life.

Letters suggest that 'The Cultivation of Christmas Trees' had been written in some form in 1952[15]; nevertheless, Eliot implied at a reading that it had been effectively composed to order, and its first draft begins truculently, declaring that he has been "bidden" to write a suitable Christmas verse.[16] The tone is that of an irritable old man, roused by professional obligation from his effective retirement to produce a poem with seasonal relevance. The 'poem' is closer in style to a didactic proclamation; there are only two sentences in its thirty-four lines, and inscriptions in copies which Eliot gifted suggest that he was unhappy with its execution, although he believed its idea to be "sound".[17] He concludes the draft by asking whether it is the kind of Christmas message one would wish to either give or receive.[18]

He reminisces now from a position of accumulated personal experience, with no need to project into an imagined old age. The memories of candles and parcels, of decorations and meals, while universal, have a descriptive particularity, along with the emotions of wonder and amazement which they engender. The child's pleasure in the sights and, more individually, the smells of Christmas echo those similarly portrayed in 'Animula' (1929), reinforcing their source in personal experience.

If we grant equal autobiographical veracity to the poem's subsequent portrayal of senescence, it suggests that Eliot's old age is proving discontented and unsatisfying. He enumerates boredom, repetition, weariness and *ennui*, together with an awareness of past failures and of future death; traits of "later experience" which merge the listless inaction of Gerontion with the painful self-criticism gifted by the compound ghost in 'Little Gidding'. And Eliot now abjures the consolation of faith, concerned that the devoutness of the convert may interfere with the children's pleasure, while being marred by self-conceit. The aridity of old age is being emphasised to contrast with the pleasure of the children; nevertheless its dysphoric character is marked.

Yet an optimism remains, that the emotional memories of Christmases past can be concentrated, and the simple wonder of the child become the awe of the adult in the face of death. The negative emotions of old age, with its awareness of mortality, are to be countered by life's possession of the amazement and joy of childhood. Once again, Eliot combines beginning and ending, the first coming and the second, young and old, and pictures an elation in both the start and the conclusion of one's life. Once again, it is a Christian spiritual perspective which grants significance to that conclusion.

As an old man, Eliot occupied a position of "exceptionally influential eminence",[19] in which his acknowledged stature gave him the authority to write tributes, obituaries and memorials to those literary figures reaching significant anniversaries or passing away. These encomia reminded the world of Eliot's own literary and social stature, associating him with eulogised individuals, and reinforcing his membership of the same lauded contingent. He also wrote eulogies and tributes to friends and colleagues who were not necessarily public figures. The increasingly frequent passing of his generation would have sharpened considerations of his own mortality; as he observed in 'Little Gidding', "we die with the dying".

In the last decade of his life, Eliot wrote some eighteen memorial tributes, two unsigned obituaries, eight tributes to living individuals for *festschrift* and birthday publications, three 'homages', and half a dozen other forms of tribute, including letters for publication in *The Times* following death announcements. Although sometimes modest in length, these compositions represent a quarter of the individual items which Eliot wrote between 1954 and his death ten years later. Their subjects reflect the three main areas in which Eliot had led his later life: poets, publishers and priests. And in summarising the essence of their subjects, they perhaps express what, as 'Little Gidding' suggests, the deceased themselves had no speech for when they were living.

For an intensely private man, who refused to authorise biographical studies, some are remarkably revealing compositions. As he writes in one of them, "In any tribute that I can pay, however brief, I find myself inevitably recording something of my own life as well".[20] They detail aspects of his life with the subjects which were not publicly revealed elsewhere, and provide an opportunity for Eliot to address the motives and deeds of the past which he had questioned in 'Little Gidding'. These divulgences, emotional as well as biographical, show old age influencing both the subjects and the content of Eliot's later public utterings.

In particular, they suggest a desire to confirm his own literary legacy. Marysa Demoor writes that "both Eliot and Pound were convinced that posthumous homage was an important aspect of literary history, in that it played a significant role in allocating an author's place in that history and in the canon".[21] Obituaries in general, she suggests, attempted "to confer immortality on the dead poet or artist [...] By doing so the author of the obituary or the memoir hoped to share in that immortality".[22]

Some, inevitably, reflect the exigencies of publishing; others, social obligation and/or a desire for association. In 1954, Eliot was invited to contribute to a *festschrift* for the seventy-fifth birthday of Wallace Stevens, a poet Eliot had only mentioned once in passing during his career, in a list of American poets.[23] Eliot clearly felt some obligation: "I write not only as an admirer, but with the special responsibility of a Director of the Firm who publish Wallace Stevens in England". He proceeds to write almost exclusively about publishing Stevens, and not about his poetry, which he says would take him too much time. "But I am only writing in order to get my name into this *Festschrift*; and I hope that the Editor will see that my name is printed so that no one can miss it".[24]

All of Eliot's pieces are essentially laudatory; such is the convention of the form, whether in celebration or commiseration. A younger Eliot might produce a piece of literary criticism rather than an encomium on such occasions, as he did in 1918 following the death of Henry James,[25] and similarly in 1930 after the death of Charles Whibley, a friend and colleague of several years, who had financially supported and contributed to the *Criterion*, and had recommended Eliot to Geoffrey Faber. Eliot wrote to Herbert Read "For Whibley I have a warm personal friendship: we have much in common".[26] Little of this is reflected however in Eliot's *Charles Whibley: A Memoir*, which Eliot describes as "attempting an estimate of the literary work of a writer whom one remembers primarily as a friend". Yet despite the unelaborated observation that "his true place in history is not altogether to be deduced by posterity merely from the writings he has left"[27] the memoir is indeed a literary consideration rather than a record of warm friendship.

But on the death of Virginia Woolf in 1941, Eliot wrote to Leonard Woolf enclosing a proof of his tribute, which focusses primarily not upon her writings but upon her cultural significance. Eliot explained that "this was not the occasion either for a critique of Virginia's work or for the expression of personal feeling".[28] In his final years, the balance on these occasions swings almost entirely from the former to the latter.

That same year Eliot wrote a furious letter to *The Times* following their obituary of James Joyce; when they did not publish it, such was his ire that he recalled the letter and had it published in *Horizon* magazine. His principal complaint was that the critical opinions of Joyce in *The Times*'s obituary were taken from "an older literary generation"—comprising Edmund Gosse, Arnold Bennett and AE (George William Russell)—and not from "Joyce's younger contemporaries, like myself".[29] This is a rare occasion on which Eliot declares himself to be "younger" than others, albeit his position is consistent with his contemporaneous denunciation in *Four Quartets* of the useless wisdom of elders. Eliot must have been angry not to have been quoted himself, given his championing of Joyce, and Faber's consequent publication of *Finnegan's Wake*. He may also have been disappointed not to have been publicly associated with such a significant literary figure. Demoor suggests that

> Eliot's frantic efforts to react to the obituary, and to get that reaction published, show the importance he attached to what was written in the obituary—not only because of what it might do to the name of a friend and esteemed colleague, but also because of what it might do to his own name, if something similar happened to him after his death.[30]

Some fifteen years later, Eliot could himself be regarded as being from "an older literary generation". He therefore confines his later tributes predominantly to his coevals, and employs modifiers such as "of *my* generation" [my emphasis] in his tribute pieces. Following the death of Wyndham Lewis, one of those contemporaries, Eliot wrote two tributes, including one for *The Sunday Times*, and then submits a further brief 'Homage' to a US student magazine, writing that: "I should not like my name to be absent from any collection of tributes".[31] This could equally demonstrate genuine respect for Lewis, obligation (to the academic Hugh Kenner, who requested Eliot's contribution), a desire for association, or a determination to maintain his literary presence. In his 'Homage' he describes Lewis as both "the most versatile prose-writer of my time" and "one of the most intelligent men of my age"[32]; both constructions allow for the potential presence of more versatile or intelligent figures emerging from a younger generation.

Those emerging figures were unlikely to achieve significant milestones, or indeed to pass away, during Eliot's remaining lifetime, so despite his status in the literary world, and his role as a publisher in seeking and promoting new talent, circumstances were less likely to arise in which he might be asked to write upon a younger poet. Instead, as an old man writing about his own contemporaries, he could simultaneously ally himself with their period of achievement, and present himself as a senior figure entitled to bestow literary status, while avoiding the framing of his judgement as that of an 'elder'.

Eliot claims in a tribute to the Italian poet Giuseppe Ungaretti[33] to place one particular constraint upon his praise of poets; he writes that he refuses to speak of any living poet as 'great'. This is consistent with an earlier critical position, stated in his essay on 'Laforgue and Corbière in our Time', that poets do not aim to be great, only to make a contribution to poetry, 'greatness' being an attribute which cannot be sought, only awarded by posterity, and dependent therefore upon things beyond the poet himself.[34] The only tribunal which can decide, Eliot writes elsewhere, is time.[35]

This could also be seen as avoiding inevitable comparisons with Eliot's own 'greatness' as a poet, establishing in public criteria against which he might shortly be measured himself. The 'time' required to endow such status could be relatively short; he described Yeats as a "great" poet several times in his lecture of 1940, eighteen months after his death.[36] He also wrote privately to Yeats's wife immediately after the poet's death, sending a message of sympathy on "the loss of the greatest living poet"[37] He identified with celerity a great editor (Bruce Richmond),[38] a great prose writer (Charles Maurras),[39] the greatest poetic dramatist of the century (Paul Claudel)[40] and the greatest man of letters of [Eliot's] generation (James Joyce).[41] In a 'Homage' of 1955, however, he *did* declare one living poet to be a "great poet"—Ezra Pound.[42]

The elderly Eliot is quick to praise when the quality of subjects' work has been maintained throughout the course of their lives. In his tribute to Virginia Woolf he considers what may be lost when a great writer dies, "unless he has already long outlived his life",[43] an unfortunately gendered remark in the circumstances, but nevertheless a striking phrase which suggests that work and life are synonymous, and that only continued literary achievement justifies continued life. Of Paul Claudel, who died at the age of eighty-six, Eliot writes, "to have been able to maintain such a standard throughout so long a lifetime is itself some evidence of greatness".[44]

He also has praise for a late renascence, observing of Edwin Muir that perhaps his finest work was written when he was over sixty, a late development reminiscent of Yeats and which represents "a triumph of the human spirit".[45] Perhaps in his own late years there is recognition here of the value placed by his grandfather upon a full lifetime of work, that "while there is life there is work to do [...] The closing years of a faithful life-service are oftentimes the best".[46] Such observations also seem to contain an acknowledgement of Eliot's own poetic silence, recognition by an old man of an achievement in their later years of others which, although he insists in 'East Coker' that there is no competition, he can no longer hope to emulate.

A particularly interesting tribute is that to Richard Aldington, a poet and writer with whom Eliot fell out of friendship. Indeed, they were "frequently at loggerheads over their writings".[47] On one occasion Eliot was upset to find himself, Vivienne and their friends lampooned in a novel by Aldington, and on another Aldington had to explain how a number of his letters from Eliot had been auctioned anonymously 'after a theft'. Mindful perhaps of the compound ghost's final gift, Eliot's tribute reads as an attempt to set down his own interpretation of events for posterity, while justifying his actions and motives. He records how "cruel and unkind" he found the lampoons, but writes that his quarrel with Aldington "subsided", implying Eliot's own magnanimity. Eliot writes that "with good intentions, but clumsy lack of imagination, I hurt his feelings once or twice very deeply indeed", accepting responsibility with a re-enactment, exploration of motives, and awareness of actions causing harm, which follows directly the sequence indicated in 'Little Gidding'. He repeats the story of his letters to Aldington seemingly in order to clarify his own understanding of the matter—although before writing his tribute he requested copies of the letters from the library which held them[48]—and he concludes the tribute in a formal and legalistic manner: "I have no reason to doubt his word and have nothing left but feelings of friendliness and regard".[49] The tribute contains no praise of Aldington's writings, nor any positive judgement upon his character beyond that single, final word. Nevertheless, Eliot's tribute appears among 22 distinguished literary contributors, including Samuel Beckett, Lawrence Durrell and C.P. Snow, in *Richard Aldington: An Intimate Portrait*, the Preface to which describes Aldington as "one of the outstanding literary figures of our time".[50] The book, containing and perpetuating Eliot's perspective, appeared shortly after Eliot's death.

Many other details of Eliot's life, career and personal relationships appear in these tributes. Although such details would be subsequently collated in posthumous biographical and critical works, these were often the first and only revelations by Eliot himself of particulars of his life. However personal his relationships with the subjects, his tributes were public—either published, or printed for congregations—and we have seen his caution that in writing them he would be "inevitably recording something of my own life as well". Given the privacy and impersonality which Eliot pursued, such revelations are therefore all the more remarkable. His tribute to Aldous Huxley, for example, recalls his life at Oxford immediately after the First World War; a visit to Garsington; an evening addressing the Poetry Circle of a ladies' club; and his first introduction to the Tarot pack.[51] His memorial tribute to Violet Schiff records the significant artistic figures he encountered in her house[52]; while that to Harriet Weaver contains his confession that he wrote his own 'Letters to the Editor' of *The Egoist* under fictitious names.[53] His 'Panegyric to Fr.

Cheetham' records his time living in ecclesiastical accommodation off the Gloucester Road.[54]

It is often in his encomia to less public figures that indications emerge of the qualities Eliot praised and respected in those of his age. These include qualities of conversation, humour, modesty and warmth of friendliness (Collin Brooks)[55]; integrity, generosity and natural goodness (Fr. Cheetham)[56]; fearlessness and independence (Arthur Stuart Duncan-Jones)[57]; and "modesty and simplicity of manner, the outward signs, I believe, of inward humility" (Bishop A.J.K. Bell).[58] These relatively simple traits illuminate the positive characteristics which Eliot acknowledged in other people of his age, countering the predominantly negative traits portrayed in elderly characters and in himself within his poetry.

Surprising perhaps is the frequency with which Eliot describes subjects as "lovable". It is an attribute he ascribes to six men and two women. In *The Cocktail Party*, Celia wonders whether it is only possible to love things created by our own imagination, and that we may all consequently be both unloving and unlovable.[59] Eliot now refutes this; his use of this simple yet intensely personal description suggests that he came to perceive and appreciate this quality in his later years. It is also an attribute which contains a concomitant admission, that Eliot felt love for these people.

The tributes provided an opportunity for such personal emotional expression, which Eliot's edicts on impersonality had often barred from his poetry. There is no more direct an expression of personal grief in Eliot's *oeuvre* than when he writes of his long-term colleague and close friend Geoffrey Faber, "I loved the man, and part of my own life is in the grave with him".[60]

Eliot wrote the first draft of *The Elder Statesman* in the autumn of 1955, when he was sixty-seven.[61] With his central character Lord Claverton a sixty-year-old, this is the only time Eliot will write an old man younger than himself. Lord Claverton has retired from a successful public career in politics and finance, in which he achieved a recognition and status in later life. The play is centred upon the relationship between this elderly public figure and a young woman, which is resolved through his confession of his past. The young woman is portrayed as his daughter, but can be seen as a proxy for Eliot's second wife, Valerie. Claverton is leading an empty life in his old age, alone without activity or purpose, and is frightened of the emptiness he sees ahead. "If I had the energy to work myself to death", he declares, in an echo of William Greenleaf Eliot,

"How gladly would I face death!"[62] No mention is made, however, of a spiritual resolution.

While the play was based upon *Oedipus at Colonus*, Eliot avoided representing the major crimes committed in Euripides. "It was more effective, for my purpose, to take some lesser crime of sin, some act of cowardice, which would be humiliating for the transgressor to think of, and which he would keep out of mind as long as he could".[63]

At the time of writing *The Elder Statesman*, Eliot was in negotiations with Emily Hale, with whom he had conducted a passionate if primarily epistolary relationship. He had told Hale in his letters that, were he not married at the time to his first wife Vivienne, he would marry her; but when Vivienne died, Eliot rescinded his proposal. Hale was now in the process of lodging Eliot's letters with a university library; Eliot was deeply concerned about their confidentiality, and negotiated a period during which they would remain sealed.

Eliot was not, he told Hale, concerned about others reading of his moments of intensity and passion; rather, as cited in Chapter 7 in relation to the "things ill-done" of 'Little Gidding', he was ashamed of "the exhibition of vanity, boastfulness, peevishness, perhaps even malice unconscious".[64] In *The Elder Statesman* Claverton defines such moments as:

> Temporary failures, irreflective aberrations,
> Reckless surrenders, unexplainable impulses,
> Moments we regret in the very next moment,
> Episodes we try to conceal from the world.[65]

These resonate with Eliot's painful re-enactments in old age expressed in 'Little Gidding'. It is not the grand Euripidean crimes of patricide and incest but these 'lesser' past actions, their motives and consequences, which cause rending pain in the elderly Eliot; and, he believes, as he described in connection with Yeats, in "most other" old men.

Out of Claverton's past, with clear echoes of Eliot's relationship with Hale, emerges a spurned fiancée, their affair recorded in letters which she has retained. "A man", she says, "is always trying to forget / His own shabby behaviour".[66] The OED defines 'shabby' as "Contemptibly mean, ungenerous, or dishonourable. Often applied, in a lighter tone, to conduct which is less friendly or generous than one had hoped for".[67]

The use of 'shabby' with its lighter tone maintains the sense of the objectively slight yet personally significant nature of Claverton's misbehaviour, while its synonym 'dishonourable' echoes the sting of the ghost's final old age gift in 'Little Gidding'.

Only by confessing his past—this affair and another modest misdemeanour—can Claverton free himself from the pretence of the public figure he has created, and allow his true self to be loved. It is hard not to speculate that during this period Eliot confessed to his own bride-to-be Valerie his prior relationship with Emily Hale, and the existence of his letters containing and then rescinding his proposal of marriage. Since the age of fourteen, Valerie had been determined to "get to Tom"[68]; her acceptance of Eliot's own "shabby behaviour" would provide reassurance to Eliot that she was in love with him, and not with his public figure. As Claverton says of Monica:

> She worshipped the part I played:
> How could I be sure that she would love the actor
> If she saw him, off the stage, without his costume and makeup
> And without his stage words.[69]

Claverton declares that he will no longer "run away" from his actions. It is hard, he explains, for people to understand how things which they regard as petty can weigh heavily upon a person. He has been accused of a morbid conscience, and told that he should have forgotten such misdeeds. But he believes that he has made mistakes, and then tried to correct them by means equally mistaken. This is the exasperated spirit of old age in 'Little Gidding', proceeding from wrong to wrong. The burden of a sin committed is not objective, Claverton says, but in relation to the sinner; it is with "the fact of my confession" that he achieves restoration:

> I've been freed from the self that pretends to be someone;
> And in becoming no one, I begin to live.
> It is worth while dying, to find out what life is.[70]

Old men who have been dying in Eliot's earlier works, like the Magus and Simeon, saw their impending death with spiritual significance. For Claverton, its value is temporal, in the late realisation of the capacity for human love. Like Eliot himself, in his end is a beginning, echoed when he reminds "my dear" that, despite his age, "I am only a beginner in the practice of loving".[71]

It is left to Monica, in the final speech of the play, to voice the transformation which has taken place in Eliot's perspective upon old age:

> Age and decrepitude can have no terrors for me,
> Loss and vicissitude cannot appal me,
> Not even death can dismay or amaze me
> Fixed in the certainty of love unchanging.
> I feel utterly secure
> In you; I am a part of you.[72]

Eliot must surely have undertaken his own confession; he certainly achieved his own resolution. On 10 January 1957, at the age of sixty-eight, Eliot married his former secretary Esmé Valerie Fletcher, who was thirty. His friend Stephen Spender, noting the reconciliation of body and soul in *The Elder Statesman*, wrote:

> In Eliot's personal life, one can rejoice that during the last ten years the synthesis was achieved, the reconciliation was complete. This fulfilment was hinted at but not realised in his poetry. Somehow, one knew all along that it could not be, that he would hymn no Yeatsian triumph of old age.[73]

Eliot was to end his years in a union which brought him a pleasure and contentment in old age that he had never envisaged.

The Elder Statesman received its first performance in 1958, with Eliot and his new young wife in the audience. The play's dedication, 'To My Wife', prefaces the published text and refers to a "book", and so is clearly intended to be part of the play's publication in print; nevertheless, it was recited from the stage before the play's first performance, a remarkably public declaration of a senior figure's passion in the presence of the couple concerned. Emended as 'A Dedication to my Wife' (1958), the subsequent work was printed in Eliot's *Collected Poems 1909–1962*, where the pagination was adjusted so that it could appear as the final work facing a verso blank page,[74] a positioning ensuring that it was perceived not as an addendum or postscript but as a final poetic epitaph. These factors underscore the significance which Eliot in his old age accorded to this uncharacteristically frank work.

In 'The Three Voices of Poetry' (1953), Eliot argues that love poems are always written with the awareness that they will be overheard—that communication to the beloved alone should be conducted in prose, presumably letters. Both versions of the dedication conclude by drawing a distinction between public statement and private content; some of the latter constituent has become clearer in recent years.

It is understandable that Eliot would have wished in his dedication to contrast the happiness of his second marriage with the misery of his first. It has also been suggested that he wanted to assert his heterosexuality, and counter the suggestion by John Peter, in an essay which Eliot had suppressed, that *The Waste Land* was based upon a homosexual relationship.[75] Having long believed that time is the ultimate arbiter of greatness, a premise he exerted in his memorial tributes to other writers, Eliot in his senescence would have been concerned about his own posthumous reputation.

Having approved Yeats's public declaration of lust in old age, Eliot in his late sixties may be similarly confirming his own vigour, particularly in a marriage to a woman thirty-eight years his younger which might have been perceived as a *mariage blanc*. In 1958, he spoke in an interview of the rejuvenation which his second marriage had brought:

> Love reciprocated is always rejuvenating. Before my marriage I was getting older. Now I feel younger at seventy than I did at sixty. Any man if he is alone becomes more aware of being lonely as he ages. An experience like mine makes all the more difference because of its contrast with the past.[76]

'A Dedication' memorialises the conjugal bliss of his final years. We have seen that previously, where sexual desires persist in Eliot's elderly characters, they lead to aberrant acts, of sexual assault, the sharing of lascivious memories, and voyeurism. Now that desire has been reinvigorated in a loving marriage, Eliot expresses a sensual passion in old age absent from his previous work. This not a conclusion to temporal life granted to any of his earlier elderly characters; it may have been unimaginable to Eliot until experienced.

The formal anonymity of the dedication's opening, "To whom", is immediately countered by Eliot's use of the first person confirming the intimately personal nature of the poem. Its expression of sensuality is unique in Eliot's *oeuvre*; spiritual passions have been replaced by sexual passion, divine love by human. The senses, which cooled in Gerontion's senescence, were cold in 'East Coker" and expiring in 'Little Gidding', are now positively quickened in Eliot's old age. Impending death, often represented by Eliot as wind and in 'A Song for Simeon' as a wind that chills, is overcome; no wind shall chill this passion. At the very end of *The*

Elder Statesman, the central characters agree that "It's chilly at dusk",[77] 'dusk' symbolising the waning in old age prior to death, but here that chill is specifically blocked.

The union has similarly triumphed in 'A Dedication' over 'withering', that physical decay in old age suffered by the Sibyl, seen in the leaves poked by the old man in 'Prufrock's Pervigilium', and in flowers, in the seasonal 'dusk' of autumn, in 'The Dry Salvages'. These roses shall not wither; the passage of time and therefore of ageing is finally being halted, and juvescence attained through a sensual relationship.

'A Dedication' suggests that love also transcends language, that the couple's private thoughts can be shared without the need for words, and that their shared speech can communicate without meaning. (In *The Elder Statesman*, the young lover Charles says "I love you to the limits of speech, and beyond".[78]) The words and meaning with which Eliot wrestles in 'East Coker' have become a 'babble', which nevertheless communicates; the poetry does not matter.

The imagery in 'A Dedication' concludes with a rose-garden, a courtly metaphor of love to be attained by a marital couple within the privacy of its enclosure. However, the rose-garden has been a significant personal symbol for Eliot, employed primarily in 'Burnt Norton', where it is one of a series of 'moments' whose profundity places them beyond the normal passage of time. Letters which Eliot wrote to Emily Hale have recently confirmed that the reference in 'Burnt Norton' refers to a visit the two made to the property together: "Our being in the rose-garden at Burnt Norton is one of the permanent moments for me".[79] Until the release of the Hale letters, this "further meaning" of the rose-garden as an aspect of Eliot's prior relationship was known only to Eliot and Hale. Its presence within 'A Dedication to my Wife' suggests therefore that Eliot had indeed confessed his relationship with Hale to his bride-to-be, just as the elderly Claverton confesses his past in *The Elder Statesman*. Eliot's declaration in his dedication to Valerie that the rose-garden is now "ours and ours only" confirms a transfer of that symbol, and a repudiation of that earlier relationship, in favour of his now wife.

The public/private dichotomy, which underpins the lines concerning speech, concludes 'A Dedication to my Wife'. These are "private words addressed to you in public",[80] shifting the addressee back to his wife; the rose-garden was being privately re-attributed, while Eliot's passion was being declared for the public and for posterity. This dichotomy can also be seen as establishing a distinction between these private poetic words, being deliberately shared with the public, and the private prose in his letters to Emily Hale being sequestered away.

Eliot's final and most personal poetic work celebrates an old age which repudiates his lifelong vision of senescence. In his own final years, Eliot seems to evade the anticipated spiritual frustrations at the prospect of death, the misuse or absence of sensual desire, the emotional pains of memories and the resentment at failing powers, all of which he had depicted during his lifetime. Instead, Eliot was to enjoy a further seven years of married happiness before his death at the age of seventy-six. One can rejoice along with Spender that, in the words of T.S. Eliot himself,

> In spite of everything, in defiance of reason,
> I have been brushed by the wing of happiness.[81]

Notes

1. Eliot, T.S., Letter to Emily Hale, 14th February 1947.
2. Hall, Donald. 1977. *Remembering Poets: Reminiscences and Opinions*. Harper & Row, 88.
3. Trevelyan, Mary and Wagner, Erica. 2022. *Mary & Mr Eliot: A Sort of Love Story*. Faber, 143.
4. Crawford, Robert. 2022, *Eliot After The Waste Land*. Jonathan Cape, 442–447.
5. https://www.npg.org.uk/collections/search/portrait/mw150379/Johann-Wolfgang-von-Goethe.
6. Eliot, T. S. 2022. *The Complete Prose of T. S. Eliot: The Critical Edition Vols 1-8*. Ed. Schuchard, Ronald, Brooker, Jewel Spears, Cuda, Anthony, Dickey, Frances, Formichelli, Jennifer, Harding, Jason, Javadi, Iman, Stayer, Jayme and Chinitz, David. The Johns Hopkins University Press and Faber & Faber Ltd. http://muse.jhu.edu/resource/1 8, 62.
7. Eliot, *Prose* 6, 83.
8. Eliot, T.S. Letter to Emily Hale, 7th March 1933.
9. Eliot, T.S. 2009-2025. *The Letters of T.S. Eliot Vols 1-10* Ed. Valerie Eliot, Hugh Haughton and John Haffenden. Faber & Faber, 10, 698.
10. T.S. Eliot, interview, *Time* Magazine, 23rd October 1950.
11. Eliot, T.S. Letter to Emily Hale, 27 Oct 1956.
12. Eliot, T.S. 1959. *The Elder Statesman*. Faber & Faber, Act Two.

13. Eliot, T.S. Letters to Emily Hale, 1st March 1947 and 20th March 1947.
14. Eliot, T.S. 2015. *The Poems of T.S. Eliot Vols I and II 1.* Ed. Christopher Ricks and Jim McCue, Faber and Faber, 1, 103.
15. Eliot, *Poems,* 1, 780.
16. Eliot, *Poems* 2, 441.
17. Eliot, *Poems* 1, 780.
18. Eliot, *Poems* 2, 442.
19. Davie, Donald. 1986. 'Virgil's Presence in Ezra Pound and Others', *Virgil in a Cultural Tradition: Essays to Celebrate the Bimillenium*, eds. Richard A. Cardwell and Janet Hamilton, University of Nottingham, 134.
20. Eliot, *Letters*, 8, 453.
21. Demoor, Marysa. 2005. 'From epitaph to obituary; the death politics of T.S. Eliot and Ezra Pound'. Biography, Spring 2005, Vol. 28, No 2 (Spring 2005), 273.
22. Ibid., 271.
23. Eliot, *Prose* 7, 805.
24. Eliot, *Prose* 8, 20.
25. Eliot, *Prose* 1, 648.
26. Eliot, *Letters* 2, 515.
27. Eliot, *Prose* 4, 384.
28. Eliot, *Letters* 9, 825.
29. Eliot, *Prose* 6, 158.
30. Demoor, 269.
31. Eliot, *Prose* 8, 174.
32. Eliot, *Prose* 8, 174.
33. Eliot, *Prose* 8, 416.
34. Eliot, *Prose* 4, 748.
35. Eliot, *Prose* 6, 577.
36. Eliot, *Prose* 6, 78.
37. Eliot, *Letters* 9, 111.
38. Eliot, *Prose* 8, 432.
39. Eliot, *Prose* 7, 773.
40. Eliot, *Prose* 8, 42.
41. Eliot, *Prose* 6, 159.
42. Eliot, *Prose* 8, 99.
43. Eliot, *Prose* 6, 169.
44. Eliot, *Prose* 8, 42.

45. Eliot, *Prose* 8, 312.
46. Eliot, William Greenleaf. 'The Personal Faith of Jesus', excerpted in Holt III, Earl K. 2011. *William Greenleaf Eliot: Conservative Radical*, Second Edition, Village Publishers, 156–7.
47. Eliot, *Prose* 8, 555, n3.
48. Eliot, *Prose* 8, 555, n6.
49. Eliot, *Prose* 8, 555.
50. Alister Kershaw & Frédéric-Jacques Temple eds. 1965. *Richard Aldington: An Intimate Portrait*. Southern Illinois University Press, viii.
51. Eliot, *Prose* 8, 560.
52. Eliot, *Prose* 8, 487.
53. Eliot, *Prose* 8, 483.
54. Eliot, *Prose* 8, 330.
55. Eliot, *Prose* 8, 336–7.
56. Eliot, *Prose* 8, 332.
57. Eliot, *Prose* 8, 145.
58. Eliot, *Prose 8*, 315.
59. Eliot, *The Cocktail Party*, Act II.
60. Eliot, *Prose 8*, 455.
61. Gordon, Lyndall. 2012. *The Imperfect Life of T.S. Eliot*. Virago, 497.
62. Eliot, *The Elder Statesman*, Act One.
63. T.S. Eliot, interview, *Grantite Review*, Vol. XXIV No. 3 (1962), 19.
64. T.S. Eliot, Letter to Emily Hale, 27th October 1956.
65. Eliot, *The Elder Statesman*, Act Three.
66. Eliot, *The Elder Statesman*, Act Two.
67. Oxford English Dictionary, "shabby (adj.), sense 2.a," December 2024, https://doi.org/10.1093/OED/1397329518.
68. Gordon, *Imperfect Life*, 490.
69. Eliot, *The Elder Statesman*, Act Three.
70. Ibid.
71. Ibid.
72. Ibid.
73. Spender, Stephen, 'Remembering Eliot', *The Sewanee Review*, Vol LXXIV No 1 Jan-Mar 1966, 84.
74. Eliot, *Poems* 2, 558.
75. Crawford, 432.

76. T.S. Eliot, interview with Henry Hewes, Saturday Review, 13 September 1958, 32.
77. Eliot, *The Elder Statesman*, Act Three.
78. Ibid.
79. Eliot, T.S. Letter to Emily Hale, 10th Sept 1935.
80. Eliot, *Poems* 1, 219.
81. Eliot, *The Elder Statesman*, Act Three.

Index

A

aberrant behaviour, 20, 32, 73, 120
adolescence as critical concept, 48–49
agèd eagle. *See* eagle (symbol)
Aldington, Richard, 134

B

Baudelaire, 67
Benson, A.C., 37, 38
blindness, 16, 18–20, 23, 24, 37–38, 68, 73, 75, 77, 85, 115
Boke named The Governour, The. *See* Elyot, Thomas

C

Callimachus, 70, 73
Cavalcanti, Guido, 90
Charlatan (in Stravinsky's Petrushka), 76, 78
Charon, 77
Coleridge, Samuel Taylor, 92
Conrad, Joseph, 77, 88
Crawford, Robert, 27

D

Dante, 31, 41, 54, 55, 58, 71, 75, 77, 90, 93, 94, 113, 118
Dream of Gerontius, The, 33

E

eagle (symbol), 91–94, 96
Eliot, Charlotte C. (mother), 14, 87, 90, 116
Eliot, T.S.
  Plays
    Sweeney Agonistes: Fragments of an Aristophanic Melodrama, 4
    The Cocktail Party, 119
    The Elder Statesman, 136–139, 141
    The Family Reunion, 66, 93, 104–106, 116, 120
  Poetry
    'A Dedication to my Wife', 36, 139–142
    'A Lyric', 17

P. Keers, *T.S. Eliot*, Modern and Contemporary Poetry and Poetics,
https://doi.org/10.1007/978-3-032-25050-6

'A Song for Simeon', 36, 40, 84, 85, 86–90, 104, 129
'Animula', 130
'Ash-Wednesday', 66, 90–95, 110, 112
'Bacchus and Ariadne: 2nd Debate between the Body and Soul', 18
'Ballade of the Fox Dinner', 20, 66
'Dans le Restaurant', 20, 31–32, 73
'Death of Saint Narcissus, The', 19–21, 24, 32, 73
'Eyes that last I saw in tears', 75
'First Debate between the Body and Soul', 18, 24
'Gerontion', 17, 20, 33–41, 66, 68, 69, 71, 75, 84, 85, 88, 95, 104, 106, 114, 119
epigraph, 19
'Goldfish (Essence of Summer Magazines)', 19
'Hysteria', 29
'In the Department Store', 17
'Journey of the Magi', 84–86
'Lines for an Old Man', 87
'Love Song of J. Alfred Prufrock, The', 21–25
'Prufrock's Pervigilium', 20, 23–24, 77, 141
Prufrock in, 19, 21–25
'Ode'(1910), 16
'Opera', 16, 96
'Rhapsody on a Windy Night', 30
'Spleen', 19
'The Cultivation of Christmas Trees', 86, 129–130
'The Hollow Men', 120
'The Little Passion', 19
'To the Class of 1905', 15
Four Quartets
'Burnt Norton', 97, 105, 112
'Dry Salvages, The', 18, 115, 117, 141
'East Coker', 104, 113–116, 117, 120, 134, 141
'Little Gidding', 17, 20, 58, 75, 98, 99, 105, 117–121, 130, 131, 138
Old Possum's Book of Practical Cats
'Old Possum persona', 96, 106–110
'Cat Morgan Introduces Himself', 111–112
'Gus: the Theatre Cat', 111–112
'Old Deuteronomy', 110–111
'The Old Gumbie Cat', 110
The Waste Land
'The Waste Land', 22, 33, 41, 64–75, 76, 93, 94
epigraph, 19, 64–65
Hieronymo, 68
manuscript, 19, 69, 88, 95
Marie, 66
'Notes on The Waste Land', 67, 69, 70
Phlebas the Phoenician, 33, 67
Sibyl of Cumae, 64–65, 67, 72, 73, 88
Tiresias, 16, 20, 41, 67, 69–74, 89
Prose
After Strange Gods, 55, 69, 99, 120
Eeldrop & Appleplex, 119
'The Defects of Kipling', 50
The Sacred Wood, 46, 64
epigraph, 34

'To Criticise the Critic', 99
'Tradition and the Individual Talent', 40, 47, 49, 52–54, 58, 70, 115
Eliot, William Greenleaf (grandfather), 14, 15, 87, 88, 116, 134, 137
'Nunc Dimittis', 87
Lectures to Young Men, 14, 39
Elyot, Thomas, 114, 116

F
Fawkes, Guy, 76
FitzGerald, Edward, 15, 37–39, 72. *See also Rubáiyát of Omar Khayyám, The*

G
Gide, André, 92
Goethe, Johann Wolfgang von, 51, 128

H
Hale, Emily, 90, 121, 127, 137, 138, 141, 142
Harris, Joel Chandler. *See Uncle Remus, The Stories of*
Hayward, John, 31, 57, 92, 118

I
immaturity as critical concept, 49–58

J
Joyce, James, 132, 134

L
Lewis, C.Day, 4, 21
Lewis, Wyndham, 21, 133

M
Maeterlinck, Maurice, 40
Mallarmé, Stéphane, 97, 98, 118
Masefield, John, 98
maturity as critical concept, 50–58

N
New Romantic Anthology, A, 121

O
Ovid, 20, 70, 73, 92

P
Pessoa, Fernando, 108
Petronius, 46, 64
Pound, Ezra, 41, 47, 65, 69, 73, 106, 108, 113, 134

R
Rubáiyát of Omar Khayyám, The, 14–17

S
Satyricon. *See* Petronius
Shelley, Percy Bysshe, 49
St Paul, 89, 93, 94
Stevenson, Robert Louis, 16, 17, 20, 25, 26, 30
Swift, Jonathan, 118
Swinburne, Algernon Charles, 49, 71
Symbolist Movement in Literature, The. *See* Symons, Arthur
Symons, Arthur, 40, 88, 97, 98

T
Tennyson, Alfred Lord, 54, 72

U
Uncle Remus, The Stories of, 108–110

V
Virgil, 41, 77, 118

W
wind as portent, 36, 68, 75, 88, 106, 117, 140
Woolf, Virginia, 3, 132, 134
Wordsworth, William, 57, 92

Y
Yeats, W.B., 4, 55–58, 113, 115, 119, 120, 128, 134
youth as critical concept, 46, 47

The manufacturer's authorised representative in the EU is Springer Nature Customer Service Centre GmbH, Europaplatz 3, 69115 Heidelberg, Germany. If you have any concerns regarding our products, please contact ProductSafety@springernature.com

Printed and bound by CPI Group (UK) Ltd, Croydon, CR0 4YY
17/09/2026
02218440-0002